NANOTECHNOLOGY SCIENCE AND TECHNOLOGY

CELLULOSE NANOCRYSTALS

ADVANCES IN RESEARCH AND APPLICATIONS

ORLENE CROTEAU
EDITOR

NOTICE TO THE READER

Library of Congress Cataloging-in-Publication Data

ISBN: 978-1-53616-747-4

Published by Nova Science Publishers, Inc. † New York

NANOTECHNOLOGY SCIENCE AND TECHNOLOGY

CELLULOSE NANOCRYSTALS

ADVANCES IN RESEARCH AND APPLICATIONS

NANOTECHNOLOGY SCIENCE AND TECHNOLOGY

Additional books and e-books in this series can be found on Nova's website under the Series tab.

CONTENTS

PREFACE

In this collection, the authors begin by presenting recent progress and preparation of cellulose nanocrystals. Additionally, different applications of cellulose nanocrystals as polymeric scaffold material, reinforcement fillers, rheology modifiers, and in electronics are reviewed and discussed.

Also provided are detailed descriptions on: cellulose nanocrystals isolation methods, characterization, and properties; current and future trends in terms of development of cellulose nanocrystals for various applications; concluding remarks on cellulose nanocrystals major uses in advanced engineering applications.

In view of the rising interdisciplinary research being carried out on cellulose nanocrystals, the authors assemble the knowledge available about the chemical structure, sources, physical and chemical procedures for the isolation of cellulose nanocrystals.

In closing, this book gives a brief, up-to-date introductory review on cellulose nanocrystals synthesis, up research in polymer-based cellulose nanocrystals nanocomposites, and their applications and challenges.

Chapter 1 - Cellulose is the most abundant biopolymer in nature which is constituted by thousands of β-D glucose units, covalently linked together by $1 \rightarrow 4$ glycosidic bonds. Owing to the hierarchical architecture of native cellulose, its nanoparticles are extracted by destructuring the native hierarchical structure. Cellulose nanocrystals (CNCs) are highly crystalline elongated rod-like (or needle-like) nanoparticles produced due to preferential dissolution of amorphous domains when cellulose substrates are subjected to a strong acid hydrolysis treatment. Because of the unique characteristics such as special morphology and geometrical dimensions, crystallinity, high mechanical strength, high specific surface area, liquid crystalline behavior, rheological properties and surface chemical reactivity etc., CNCs have several potential applications as polymeric matrix materials, reinforcement fillers, rheology modifiers, and liquid crystalline materials for liquid crystalline displays. Therefore, CNCs have garnered a great deal of research interests focused on the optimization of preparation processes, tailoring properties for a specific application, and exploring new avenues of applications. In the first section of this chapter, the authors present recent progress and preparation of CNCs. In the second section, different applications of CNCs as polymeric scaffold material, reinforcement fillers, rheology modifiers, and in electronics are reviewed and discussed.

Chapter 2 - In the 21st century, cellulose based nano-sized materials will be regarded as an innovative matter that encouraging production of high value-added pulp and paper products, and provide solutions for technological innovations for various applications. Recently, there is a wide range of cellulose based nano-materials that have been extracted, and developed for various applications. The current research interest is towards conducting a large scale research and development activities on numerous applications of nano-materials for commerical use. Cellulose nanofiber is a eco-friendly materials with a potential applications as reinforced filler in various composites, targeted drug delivery, smart materials, 3D printing, and automotive interior parts. The first cellulose nanofiber or micelles was chemically (acid hydrolysis) extracted by Ranby in 1949. Based on fabrication techniques, cellulose nanofibers can be categorized into cellulose nanocrystals, nanofibrillated celulose and TEMPO cellulose nanofibers. Cellulose nanofibers can be extracted by top down and bottom up approaches such as mechanical, enzymatic hydrolysis, TEMPO, acid hydrolysis and production of bacterial cellulose respectively. In recent years, the use of cellulose nanocrystals for different applications have gained much research interest because they are bio-degradable, bio-compatible, carbon neutral, and readily available from renewable resources. In the acid hydrolysis process, the cellulose nanocrystals (CNCs) amorphous regions are destroyed while the crystalline domains are kept intact. In this review chapter, the recent developments on research and applications of cellulose nanocrystal extracted by acid hydrolysis will be discussed. This chapter aims to provide a detailed description on i) Cellulose nanocrystals isolation methods, characterization, and properties ii) Current and future trends in terms of development of CNCs for various applications and iii) Concluding remarks on cellulose nanocrystals major use in advanced engineering applications such as composites, smart materials, electronic devices, automobile interiors, surface coating, used as a lubricant, transparent films, targeted drug delivery and etc.

Chapter 3 - Cellulose nanocrystals (CNCs) are exclusive nanomaterials (NMs) derived from the most abundant and almost inexhaustible natural polymer. These NMs have gathered the attention of the scientific community due to their unique mechanical, optical, chemical, and rheological properties. CNCs are biodegradable, biocompatible and renewable, hence serving as a sustainable and environmentally friendly material as they are mainly obtained from naturally occurring cellulose fibers. In view of the rising interdisciplinary research being carried out on CNCs, this review aims to assemble the knowledge available about the chemical structure, sources, physical and chemical procedures for the isolation of CNCs. The description of the mechanical, optical, and rheological properties of CNCs is also given in this review. Innovative applications in diverse fields such as pharmaceutics, catalysis, food, and packaging have also been discussed. In addition, latest advances in the fields of wound healing, regenerative medicines and drug delivery have also been highlighted.

Chapter 4 - Cellulose is the most abundant biopolymer on earth has been applied in almost all areas/field of material science and engineering with cellulose nanocrystals (CNC) being one of the highly researched/applied form of cellulose recently. This book chapter gives a brief up to date introductory review on CNC, its synthesis approach, up to date research in polymer-based CNC nanocomposites, its applications, challenges and scope for future applications and conclusion. The chapter serves as a guide for researchers, scientist, engineers, and technologist to know areas where there is still room for utilization of CNC in advance materials applications and also to effectively develop/proffer new novel application areas for CNC.

ABBREVIATIONS

AgNPs	Silver nanoparticles
AuNPs	Gold nanoparticles
BC	Bacterial cellulose
BC-Ch	Bacterial cellulose-Chitosan
CDAP	1-cyano-4-dimethylaminopyridinium
CNCs	Cellulose nanocrystals
CNFs	Cellulose nanofibrils
CuNPs	Copper nanoparticles
HDFs	Human dermal fibroblasts
MMP	2-methoxy-4-methylphenol
NCC	Nanocrystalline cellulose
NFC	Nanofibrillated cellulose
NMs	Nanomaterials
PdNPs	Palladium nanoparticles
PS	Potato starch
PtNPs	Platinum nanoparticles
RuNPs	Ruthenium nanoparticles
TEMPO	(2, 2, 6, 6-Tetramethylpiperidin-1-yl) oxyl radical
UDPGlc	Uridine diphosphoglucose
WVTR	Water vapour transmission rate
4-AP	4-Aminophenol
4-NP	4-Nitrophenol

In: Cellulose Nanocrystals
Editor: Orlene Croteau

ISBN: 978-1-53616-747-4

Chapter 1

CELLULOSE NANOCRYSTALS - SOURCES, PREPARATION, AND APPLICATIONS: RESEARCH ADVANCES

Sanjit Acharya, Shaida Rumi Sultana, Prakash Parajuli and Noureddine Abidi*
Fiber and Biopolymer Research Institute,
Department of Plant and Soil Science,
Texas Tech University, Lubbock, TX, US

ABSTRACT

Cellulose is the most abundant biopolymer in nature which is constituted by thousands of β-D glucose units, covalently linked together by $1 \rightarrow 4$ glycosidic bonds. Owing to the hierarchical architecture of native cellulose, its nanoparticles are extracted by destructuring the native hierarchical structure. Cellulose nanocrystals (CNCs) are highly crystalline elongated rod-like (or needle-like) nanoparticles produced due to preferential dissolution of amorphous domains when cellulose substrates are subjected to a strong acid hydrolysis treatment. Because of the unique

*Corresponding Author's Email: noureddine.abidi@ttu.edu.

characteristics such as special morphology and geometrical dimensions, crystallinity, high mechanical strength, high specific surface area, liquid crystalline behavior, rheological properties and surface chemical reactivity etc., CNCs have several potential applications as polymeric matrix materials, reinforcement fillers, rheology modifiers, and liquid crystalline materials for liquid crystalline displays. Therefore, CNCs have garnered a great deal of research interests focused on the optimization of preparation processes, tailoring properties for a specific application, and exploring new avenues of applications. In the first section of this chapter, we present recent progress and preparation of CNCs. In the second section, different applications of CNCs as polymeric scaffold material, reinforcement fillers, rheology modifiers, and in electronics are reviewed and discussed.

1. INTRODUCTION

Cellulose is the most abundant biopolymer on earth. It is biosynthesized by living organisms (plants, some species of bacteria, algae and tunicates- the only known animals capable of biosynthesizing cellulose) as a linear homopolymer of thousands of repeating β-D glucose units, covalently linked together by polycondensation reactions between hydroxyl groups at C1 of a glucose unit and C4 of the neighboring glucose unit (Figure 1). Dimer of glucose, commonly known as cellobiose, is considered as a repeating unit of cellulose polymer because every repeating unit is rotated 180° with respect to its neighbors along the fiber axis [1]. However, debate regarding whether or not cellobiose is the repeating unit has resurfaced and a recent review in this regard makes a cogent argument that glucose is the repeating unit rather than the cellobiose [2]. Cellulose is ubiquitous in plants and is generally a fibrous, tough and water-insoluble material, which contributes vitally to the structural integrity of the plant cell walls [3].

Cellulose in nature is not found as an isolated molecule but rather organized into an intriguing multi-level assembly, known as hierarchical structure [4]. According to the usually agreed model, for plants, an elementary fibril is comprised of 36 cellulose chains and has a square cross-section (3-5 nm in size). These elementary fibrils further assemble to a larger dimensional microfibril. Ultimately, several microfibrils are bundled together to form a native cellulose fiber. These microfibrils further assemble into macro sized cellulose fibers (e.g., wood fiber, cotton fiber) [1, 3]. However, specific packing of cellulose might be different based on the source.

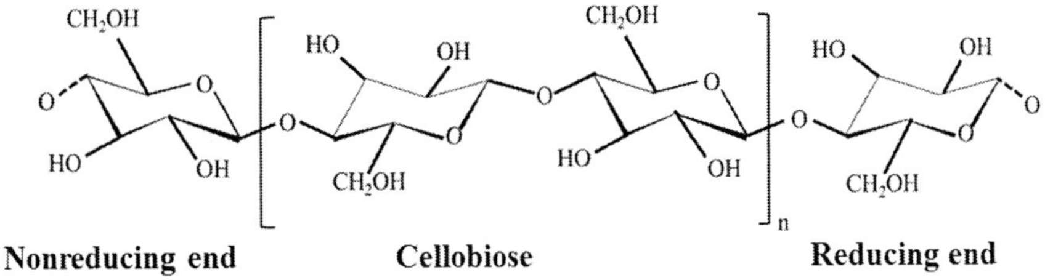

Nonreducing end **Cellobiose** **Reducing end**

Figure 1. Chemical structure of cellulose.

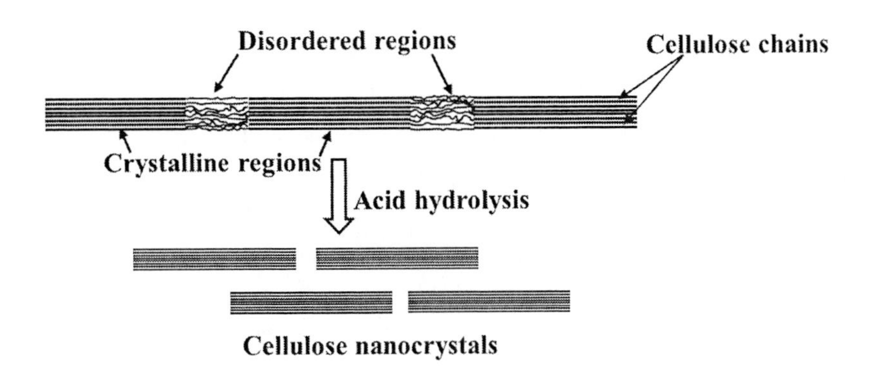

Figure 2. Schematic of cellulose microfibril showing distribution of crystalline and amorphous regions and cellulose nanocrystals.

Fundamental assemblies of cellulose, cellulose fibrils act as the main reinforcement phase for plants, algae, marine creature tunicates and some species of bacteria, which secrete cellulose fibrils creating extra-cellular network structure. These cellulose fibrils consist of crystalline (where cellulose chains are arranged in a highly ordered structure) and amorphous regions (where arrangement of cellulose chains is disordered) (Figure 2). However, the structure and distribution of crystalline and amorphous regions are not fully resolved yet [5].

Fascinating hierarchical architecture of native cellulose allows us to extract nanoparticles from this most abundant, virtually inexhaustible, renewable natural polymer. Thus, the term "nanocellulose" encompasses various materials derived from native cellulose (found in plants, animal-tunicates, algae, and bacteria), which possess at least one dimension in the nanometer range [3, 6]. Nanocellulose possesses properties, which are different from conventional materials. Cellulose nanoparticles feature unique characteristics such as special morphology and geometrical dimensions, crystallinity, high specific surface area, liquid crystalline behavior, alignment and orientation, rheological properties, mechanical reinforcement, barrier properties, surface chemical reactivity etc. Moreover, other beneficial properties such as biocompatibility, biodegradability, low toxicity coupled with abundance and renewability (nanocellulose is extracted from cellulose) make nanocellulose capable of offering new horizons in nanotechnology [7, 8]. The extraction of cellulose in nanoscales enables us to remove most of the defects associated with the structures in the hierarchy of the native cellulose rendering impressive mechanical properties and reinforcing capabilities. With Young's modulus in the range of 100-140 GPa and a surface area of several hundreds (m^2g^{-1}), new promising properties can be materialized for cellulose based materials [3, 9–11].

In general, nanocelluloses are grouped into three broad categories: 1) cellulose nanofibrils (CNF), 2) cellulose nanocrystals (CNCs), and 3) bacterial cellulose (BC). Generally, multiple mechanical shearing actions are employed to the already purified cellulose substrate to obtain enmeshed individual or bundle of fibrils called CNFs. CNFs contain both amorphous and crystalline regions. On the contrary, when cellulose substrates are subjected to a strong acid hydrolysis treatment, preferential dissolution of amorphous domains results in highly crystalline elongated rod like (or needle-like) nanoparticles called CNCs. Unlike the production of CNFs and CNCs, bacterial cellulose is secreted by certain species of bacteria (e.g., *Acetobacter xylinum*) in the form of microfibrils, which constitute extracellular matrix. These microfibrils further aggregate as ribbons ultimately generating web-shaped network structure of cellulosic fibers.

Typical sources, the method of production, and the average size of three general categories of nanocelluloses are summarized in Table 1.

Table 1. General categories of nanocellulose [12]

Nanocellulose type	Synonyms	Typical sources	Typical methods of production and average size
Cellulose nanofibrils (CNF)**	Nanofibrillated cellulose (NFC), nanofibrils and microfibrils, cellulose microfibrils, microfibrillated cellulose (MFC)	Wood, sugar beet, potato tuber, hemp, flax etc.	Transverse cleavage along the longitudinal direction of cellulose microfibrillar structure by multiple mechanical shearing actions before and/or after chemical or enzymatic treatments of cellulose to release more or fewer individual microfibrills Diameter: 5-60 nm Length: several microns
Cellulose nanocrystals (CNC)	Nanocrystalline cellulose (NCC), cellulose crystallites, cellulose (nano) whiskers, rod-like cellulose microcrystals	Wood, cotton, hemp, flax, wheat straw, mulberry bark, ramie, microcrystalline cellulose (MCC), Avicel, tunicates, algae, bacteria etc.	Preferential removal of amorphous regions in the cellulose fibrils by acid hydrolysis Diameter: 5-70 nm Length: 100-250 nm (from plant); 100 nm-several microns (from cellulose of tunicates, algae and bacteria)
Bacterial Cellulose (BC)	Bacterial nanocellulose (BNC), microbial cellulose, biocellulose	Low molecular weight sugars and alcohols	Bacterial synthesis Diameter: 20-100 nm; different types of nanofiber networks

** Often times, cellulose nanofibrils and cellulose microfibrils are interchangeably used. However, cellulose nanofibrils have finer particle diameters (4-20 nm) than that of cellulose microfibrils (10-100 nm) [3].

Although all three classes of nanocellulose have attracted great deal of both fundamental and applied research interests in the field of nanotechnology, we focus on cellulose nanocrystals (CNCs) in this book chapter.

2. SOURCES OF CNCS

Since CNCs are derived using "top-down" approaches from native cellulose fibers, a variety of biological species, which can synthesize cellulose can serve as sources of CNCs. Therefore, different types of sources such as plant cell walls, cotton fibers, algae, bacteria, and animals (tunicates) can be exploited to derive CNCs [1, 3]. However, the resulting CNCs differ in morphology, structure, functional properties and potential applications depending on the source, the origin, the maturity, and other processing parameters [13, 14]. For example, CNCs derived from woody biomass using acid hydrolysis are usually spherical or shorter rod-shaped with lower aspect ratios (10-30), while CNCs obtained from bacteria and tunicates, typically, have higher aspect ratios (aspect ratio ~ 70, several micrometers in length for tunicate CNCs) [15]. In the case of sulfuric acid hydrolyzed CNCs, the effect of hydrolysis conditions on their thermal stability is well pronounced. The thermal stability of CNCs is enhanced when very short hydrolysis time is used. Yet, reducing the hydrolysis time results in larger CNCs with lower colloidal stability [16].

2.1. Lignocellulosic Sources

As plants are the predominant sources of cellulose and cellulose in plants constitutes approximately one-third of all plant materials [17], lignocellulosic biomass is the obvious raw material to produce cellulosic fibers and ultimately CNCs. Both woody and non-woody plants can be considered as natural biocomposites ingeniously crafted by nature, where cellulose microfibrils are embedded in the matrix of hemicellulose/lignin, waxes/pectin, and other trace materials [18]. Pure cellulose can be obtained after removal of these matrix materials from the plant biomass. Currently, wood and/or wood-based cellulose is the most important raw material for producing CNCs both in laboratory and industry although nanocellulose industry is still in nascent stage [15]. Besides wood, by virtue of being the purest source of cellulose in nature, cotton fiber is another important and conventional source of CNCs [19, 20]. Apart from conventional sources, recently, a wide array of annual plants, other nonconventional plant species, agricultural waste, and agro-industrial waste have been investigated for extraction of CNCs. The list includes, but not limited to, alpha fibers [11], flax fibers [21], hemp fibers [22], sugarcane bagasse [23–25], cassava bagasse [26], wood sawdust [27], maize stover [28], banana pseudostem [29], cornhusk [30], corncob [31], coconut husk fiber [32], rice husk [33–35], sago seed shells [36], garlic straw [37], onion skin waste [38], grain straws (rice, wheat, barley) [39], waste cotton cloths [40], sesame husk [41], passion fruit peels [42], grape skins [43], tomato peels [44], potato peels [45], pomelo albedo [46], pineapple leaf [47], mango seed [48], pistachio shells [49], groundnut shells [50], bamboo fiber [51, 52], *Miscanthus x. Giganteus grass* [53], blue agave waste (leaves and bagasse) [54], screw pine (*Pandanus tectorius*) leaves [55], mulberry barks [56], and chili leftovers [57]. Recent growing interest in nonconventional sources (herbaceous plants, agricultural crops and their byproducts) for extraction of CNCs might be due to the realization that, with the growing demand of CNCs (nanocellulose) in the future, sole dependency on wood-based cellulose might not be a wise strategy because wood might not be available in sufficient quantities at reasonable price due to competition among various sectors viz. building construction, furniture, and pulp and paper industries. The extraction and purification processes of cellulose from non woody plants are comparatively less harsh and less energy demanding as compared to wood because the latter typically contains more lignin than the former ones [15]. Moreover, the utilization of agricultural waste/byproducts in preparation of CNCs offers new economic opportunities from large amount of waste, which is almost unutilized until today and, in some cases, could be a pernicious environmental menace especially when burnt generating air pollution [58].

2.2. Bacterial Sources

Cellulose secreted by bacteria as extracellular matrix, also known as bacterial cellulose (BC) is receiving tremendous attention because of its high purity (free from hemicellulose and lignin), which makes it suitable for uses to produce fibers and extract bacterial cellulose nanocrystals (BCNCs). Several species of bacteria such as *Komagataeibacter xylinus* (previously known as *Gluconacetobacter xylinus or Acetobacter xylinum*) [59], *Agrobacterium*, *Rhizobium*, and *Sarcina* [60]. *Komagateibacter xylinus*, gram-negative acetic acid bacteria is the most studied species for BC production and is considered as the most efficient source of BC known so far [15, 59]. Bacterial cellulose also features other desirable properties such as high porosity, high surface area, high water holding capacity, high crystallinity, high mechanical strength, and low toxicity [61]. Salient properties of BC, especially high purity and high crystallinity make it an excellent starting material for producing CNCs. However, low production rate and high production cost of BC seriously limit its practical applications [62]. Nonetheless, BCNCs may still be feasible for specific applications particularly in pharmacological and biomedical fields [61, 63].

2.3. Algae Source

Another non-plant cellulose source is algae. Different types of algae such as green, gray, yellow-green, red, etc. produce cellulose microfibrils. Algae is an informal designation of a diverse group of photosynthetic organisms, which are not necessarily evolutionarily closely related. A wide range of variability in cellulose microfibrillar structure has been observed due to differences in biosynthesis process [64]. Green algae such as *Valonia* and *Cladophora* are well known for highly crystalline cellulose. The crystallinity of cellulose from *Valonia* and *Cladophora* is >95%. Algal cellulose, originating from various algae species (both marine and freshwater) has been utilized for the preparation of CNCs. Produced CNCs have also been used in various polymeric nanocomposites [65, 66].

However, because of the low cellulose content, and high cost of collection and extraction, algae are not considered as a viable source of CNCs for large-scale production. Nonetheless, there is some optimism that red algae such as *Geldium*, which are rich in carbohydrate content, could be new promising candidates for preparing CNCs (cellulose nanomaterials) because of their abundance and availability [15].

2.4. Animal Source

Tunicates, the marine invertebrate animals, are the only known source of cellulose from animal kingdom so far. Cellulose microfibrils are embedded in a protein matrix in the outer tissues called tunic [3]. Similar to cellulose from plants, microfibrils in tunicate cellulose (TC) have, almost exclusively, crystalline arrangement of cellulose I_{β} [67]. Due to the high aspect ratio of tunicate cellulose microfibrils, there is a growing interest in tunicates as a source of high aspect ratio tunicate cellulose nanocrystals (tCNCs). Although some studies have shown otherwise [68], it is, generally, well established that CNCs extracted from tunicates have the highest aspect ratio among CNCs from all other sources [3]. Higher aspect ratio and stiffness of tCNCs, compared to CNCs from other sources, make them superior reinforcing fillers. High aspect ratio and high Young's modulus of tCNCs contribute to the onset of percolating networks and stress transfer at lower filler density, and higher stiffness of composite polymer matrices containing tCNCs respectively [53, 67]. Among three classes of tunicates, only two classes namely Ascidiacea and Thaliacea contain tunics, the tissues containing cellulose. *Halocynthia roretzi*, *Ascidia sp.*, *Ciona intestinalis*, *Halocynthia papillosa*, *Mentandoxarpa uedai*, and *Styela plicata* have been some of the most frequently studied species. Despite superior attributes of tCNCs and their importance as idealized particles for research purposes, there is well-founded skepticism, in the scientific community in cellulose nanomaterials field [53, 62], about tunicates as a viable source of CNCs for commercial-scale production because of the high cost associated with the collection of tunicates and subsequent extraction of cellulose from them.

However, tunicate species such as *Ciona intestinalis*, which could be reared at very high densities in the ocean with potential of large scale cellulose fabrication, might be a promising source of CNCs [15].

3. Method of Extraction of CNCs

The process of extraction of CNCs from cellulose source materials typically involves two major steps: pretreatment and hydrolysis to remove the amorphous regions. Multiple pretreatment steps are generally required for purification/isolation, and homogenization of cellulose from raw source materials unless pure cellulose sources such as bleached wood pulp, cotton fibers, and microcrystalline cellulose (MCC) are used [15]. Generally, pretreatments are followed by subjecting purified and homogenized cellulose substrate to controlled chemical treatment (most typically hydrolyzed with sulfuric acid). Post-hydrolysis steps involve centrifugation and dialysis, and sonication to separate, neutralize, and disaggregate the extracted CNCs respectively. Figure 3, schematically summarizes the general process of obtaining CNCs from raw cellulose. Specific processes in the extraction of CNCs (both pretreatments and subsequent hydrolysis) affect yield, dimension, homogeneity, and perfections.

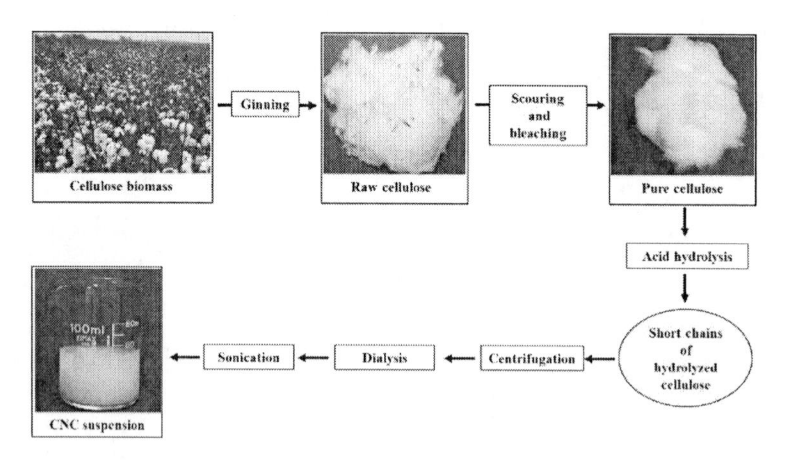

Figure 3. Schematic of preparation of CNCs from cellulosic biomass.

3.1. Pretreatment of Raw Cellulose Sources

The primary purpose of the pretreatments is to completely or partially make cellulose free from matrix materials (e.g., lignin and hemicellulose in wood and plants, bacterial cells and proteins in the case of bacterial cellulose). The objective of the pretreatments is also to isolate individual complete fibers (e.g., from woods) or fibrils (e.g., from tunicates). Specific pretreatment is employed based on the source of cellulose material and the desired morphology of the cellulose particles to be used in the subsequent treatments [3]. In the following discussion, the general pretreatment processes are briefly discussed with a focus on lignocellulosic biomass.

For lignocellulosic biomass, the pretreatment process begins with the reduction of size of raw source material into smaller fragments by milling/grinding and cutting. Often times, raw lignocellulosic materials are milled into powder form, which makes it easier to handle and also increases their contact surface area during subsequent scouring (alkali treatment) and bleaching treatments, which are necessary in order to remove noncellulosic matrix materials such as lignin and hemicellulose. Although washing raw cellulose source material can be judiciously skipped for some fibers, which do not contain significant amount of dirt such as sugarcane bagasse and cotton fiber, washing is usually necessary. Ground lignocellulosic raw materials are washed by dispersing or soaking them in water under constant stirring followed by filtering. This washing and filtering step also helps in removing water-soluble extractives and impurities and also in dewaxing. Dewaxing can be performed in a Soxhlet apparatus with a 2:1 (v:v) mixture of benzene and ethanol [69, 70].

Ground and washed raw fibers are then subjected to alkali treatment using aqueous sodium hydroxide (NaOH)/potassium hydroxide (KOH) solution. Alkali treatment removes alkali-soluble materials like hemicellulose and other impurities. At least, 50% removal of hemicellulose has been reported to be required to increase the cellulose digestibility. Broadly, two types of alkali treatments can be employed. One involves the heating of raw cellulose fibers in alkali solution at 70 – 90°C combined with mechanical stirring. Another method of alkali treatment is performed by first cooling the fibers followed by high temperature treatment at 160°C. Alkali treatment has been reported to be capable of removing most of the hemicellulose leading to destruction of raw fiber structure and exposure of the surface with fewer impurities.

Bleaching is the next step carried out on alkali treated cellulose source material to ensure the removal of remnant non-cellulosic materials especially lignin. Therefore, bleaching is also sometimes termed as delignification. Since lignin is the hardest noncellulosic material to remove, and higher amount of lignin seriously interferes with the reactivity of cellulose during the hydrolysis step, keeping the amount of remnant lignin as low as possible is extremely important. Although sodium chlorite ($NaClO_2$) has been popular as a bleaching agent [71], other reagents for bleaching purpose have been sodium hypochlorite (NaOCl) and hydrogen peroxide (H_2O_2). In a typical bleaching process, alkali treated fibers are boiled in $NaClO_2$ solution under acidic condition. Relatively constant acidic condition ~ pH 4 is maintained by using acetate buffer solution, prepared using glacial acetic acid and sodium acetate. Chlorine dioxide (ClO_2), breakdown product of $NaClO_2$ in acidic buffer solution oxidizes lignin leading to further purification of cellulose fibers [72]. Often times, the bleaching process is repeated several times until the fibers become white indicating the removal of noncellulosic impurities to higher degree [73]. Higher amount of remnant lignin not only impedes the hydrolysis by restricting the easier access of acid to the inside of the cellulose fibrillar structure but also adds to the imperfection of the resultant CNCs (e.g., decreased crystallinity) [71].

Even though efficient pretreatments of lignocellulosic biomass are indispensable in the preparation of CNCs, there is a growing concern about the cost and environmental impacts of the conventional alkali and bleaching treatments. Bleaching is especially problematic from the environmental perspective as compounds such as oxidized lignin, carbohydrates, and some organochlorines are inevitably released with the effluent. These compounds are later broken down into much nefarious long life dioxins and furans. Furthermore, concentrated fumes of ClO_2 are toxic to humans. These concerns have prompted researchers to look for relatively benign pretreatment methods. Shin et al. reported that alkali treatment step can be waived when raw fibers were subjected to electron irradiation beam followed by cooking in water without any chemicals [74]. Recently, another study claimed to separate cellulose fibers using "fully-environmentally friendly" pretreatments [54]. In the study, firstly, raw biomass from blue agave was subjected to an ethanol-water (70:30, v/v) organosolv treatment at 200°C and pressure ~30 bar for 90 min. Then, the chlorine free bleaching of treated fibers was carried out using oxygen and hydrogen peroxide in alkaline aqueous solution [54]. Other alternatives, to the conventional pretreatments, have been biological/enzymatic pretreatments of raw cellulose biomass. Certain enzymes such as xylanase and pectinase can attack hemicellulose and alter the original interface between lignin and cellulose, which facilitates the removal of the lignin associated with hemicellulose fraction [15, 22].

With the exception of bacterial source, pretreatment methods for non-plant sources (tunicates and algae) are also similar in a sense that raw fibers are subjected to alkali and bleaching treatments. However, specific treatment parameters vary. On the other hand, for bacterial cellulose, no harsh chemical pretreatments are required unlike lignocellulosics to remove hemicellulose and lignin as BC is pure cellulose. Washing with dilute aqueous alkali is sufficient to remove impurities (bacterial cells, ingredients from medium, etc.) [15].

3.2. Preparation of Cellulose Nanocrystals (CNCs)

The isolation of CNCs from purified cellulose fibers is a crucial stage in the extraction process of CNCs. CNCs are isolated by selective removal of the amorphous domains (regions of disordered fibrillar organization) in the cellulose fibrillar structure. Although cellulose source and pretreatments also affect the properties of liberated nanocrystals (CNCs) [75], specific parameters selected during the removal of the amorphous domains largely dictate the properties of liberated CNCs [76]. Therefore, selecting wisely the parameters and controlling the CNCs liberation process is of paramount importance as their superior properties are crucial for further processing into high-value products (composites). Furthermore, efficiency, cost, and environmental issues also need to be taken into consideration. Several methods of extraction of CNCs such as acid hydrolysis, enzymatic hydrolysis, ionic liquid treatment, oxidative process, subcritical water hydrolysis, mechanical refining, and combined processes have been reported. Nonetheless, acid hydrolysis has remained the most widely employed method of isolation of CNCs, and among several acids used, sulfuric acid hydrolysis is the most popular method both in academia and industry [16]. Studies are focused on the optimization of the process to achieve narrow size distribution of CNCs as well as their tailorable charge content, high yield, thermal stability and crystallinity [16]. In this section, a brief overview of the various CNC's isolation methods is presented.

3.2.1. Acid Hydrolysis

Acid hydrolysis is the most widely employed method for the isolation of CNCs from cellulose fibers. When purified cellulose substrate is treated with preferred acid at strict reaction conditions (acid concentration, agitation, time and temperature), owing to the differences in kinetics of hydrolysis between amorphous and crystalline domains, preferential hydrolysis of cellulose chains in the amorphous domains of cellulose microfibrils along the fiber axis occurs. Consequently, the disintegration of the hierarchical structure of fiber bundles leads to the release of individual CNCs as illustrated in Figure 2. The acid hydrolysis process begins with the diffusion of hydronium ions, released from the acid, inside the amorphous domains and subsequent protonation of oxygen species on the glycosidic bonds, thereby leading to the hydrolytic cleavage of the glycosidic bonds of the amorphous regions. Residual pectin and hemicellulose could also be hydrolyzed into simple sugars by the acid during the acid hydrolysis reaction. During the acid hydrolysis process, abrupt decrease in the degree of polymerization (DP) of cellulose substrate occurs until a relatively constant DP termed levelling-off DP (LODP) is reached. The DP continues to decrease even after the LODP is reached because the crystalline sheets are also prone to be peeled-off from the crystallites. However, the process is quite slower even during prolonged hydrolysis time. The LODP is thought to correlate with the size of the crystals along the longitudinal direction of cellulose fibers prior to acid hydrolysis. After being liberated, the original crystallites can grow in size, and consequently possess dimensions larger than the original microfibrils, due to the availability of the large freedom of motion after hydrolytic cleavage [77]. Once the hydrolysis reaction is complete or quenched usually by adding large excess water to the reaction mixture, CNCs are separated from the reaction mixture by several rounds of centrifugation combined with washing. Separated CNCs are dialyzed against deionized water to remove residual acid and neutralized salts. Finally, in order to facilitate disaggregation and homogenous dispersions of CNCs in aqueous media, sonication treatment is usually applied.

Even though, sulfuric acid and hydrochloric acid are frequently used to carry out the hydrolysis of cellulose substrates for the preparation of CNCs, several other strong as well as weak acids such as hydrobromic acid, nitric acid, phosphoric acid, formic acid [78], acetic acid, citric acid, oxalic acid, maleic acid, hydrogen peroxide, and a mixture of acids [27] have been employed for this purpose. Standard sulfuric acid hydrolysis process for CNCs preparation constitutes 64 wt% acid concentration at 45–50°C for approximately 60 min. Nonetheless, the reaction parameters might be slightly varied. The reaction conditions (acid employed, acid concentration, temperature, acid to pulp ratio (acid to solid ratio), and hydrolysis time) have profound effects on the structural (morphology, crystallinity, and aspect ratio) as well as functional (thermal stability, colloidal stability, etc.) properties of CNCs. For example, Zhang et al. prepared CNCs from isolated bamboo cellulose using four different acids, namely: sulfuric, hydrochloric, phosphoric acids, and a mixture of acetic and nitric acids [52]. The study showed that the resultant CNCs had the length of 3–200 nm (using sulfuric acid), 20–85 nm (using hydrochloric acid), 20–40 nm (using phosphoric acid), and 6.5–20 nm (using mixture of acetic acid and nitric acid). The study also showed that the nanocrystals prepared using sulfuric acid and phosphoric acid possessed higher crystallinity and lower thermal stability compared to those prepared using hydrochloric acid and a mixture of acetic acid and nitric acid [52]. Yield is another important parameter to consider for commercial-scale production. Usually, harsher reaction conditions (e.g., increasing hydrolysis time, temperature, acid concentration, and acid to pulp ratio) result in smaller CNCs with higher yield. However, further increase in the harshness of the hydrolysis reaction leads to lower CNC yield and decreased crystallinity due to the degradation of cellulose to sugars. On the other hand, mild reaction conditions (e.g., decreasing hydrolysis time, temperature, acid concentration, and acid to pulp ratio) lead to larger CNCs but the yield is low due to insufficient hydrolysis resulting in higher amount of partially hydrolyzed fibers in the form of precipitable cellulose solid residue (PCR) [79]. A typical sulfuric acid hydrolysis process yields ~30% of CNCs. Therefore, the reaction parameters can be judiciously varied to obtain cellulose nanocrystals with different properties based on the targeted applications. Table 2 summarizes the properties of CNCs extracted from different starting materials using acid hydrolysis under different reaction conditions.

3.2.2. Enzymatic/Microbial Hydrolysis

Several studies have shown that the production of nanosized cellulose/CNCs can be realized with enzymatic/microbial hydrolysis [104–106]. Endo-β-1,4-glucanase (endoglucanase, EG), one of the components of the multi-component cellulase enzyme system can be used to synthesize CNCs as EG predominantly acts on the amorphous regions of cellulose and randomly hydrolyzes β-1,4 glycosidic bonds shortening the original cellulose chain length. However, when using enzymatic hydrolysis of cellulose for the production of CNCs, activities of other two components of cellulase enzyme system, namely viz: *exo*-β-1,4-glucanase (exoglucanase, CBH) and β-glucosidase (GB) should be inhibited. Exoglucanase acts on both ends (reducing and non-reducing) of cellulose to hydrolyze the glycosidic bond and cleave cellulose from the end in cellobiose. β-glucosidase (GB) finally hydrolyzes soluble oligo-glucose, produced after the action of EG and CBH, to glucose [107, 108].

Table 2. Morphologies and yield of CNCs from different acid hydrolysis processes

Acid used	Hydrolysis parameters	Cellulose source	CNCs size	Yield (wt%)	References
	Acid concentration/ acid:solid ratio/temperature/ time (wt%/v:m/°C/min)		Length/width (nm/nm)		
Sulfuric acid	32/10:1/50/300	Cotton fibers	450/25	-	[80]
	52/10:1/50/60	Passion fruit peels	-	58.1 ± 1.7	[42]
	46.86/20:1/55/300	MCC from cotton	75.74 ± 5.25/10.31 ± 1.54	76.33	[81]
	65/40:1/55/60 mins	Cotton gin motes	122 ± 79/6.9 ± 5.6	54.5	[82]
	63.5/9:1/45/90	Cotton fibers	131.4 ± 64.8/50.5 ± 6.6	-	[20]
	64/-/45/30	North African grass (*Ampelodesmos mauritanicus*)	180 ± 40/21.2 ± 5.1	8.7	[70]
	55/-/65/150	Blanched cellulose pulp	100 – 400/7 – 10	85	[72]
	62/8:1/50/70	Bleached Kraft pulp	204 ± 129/17.3 ± 16.1	59.7	[83]
	65/20:1/45/40	Garlic straw	480/6	19.6	[37]
	55 (v/v)/-/45/20	Tunicate cellulose	694 ± 312/20 ± 2.8	30	[84]
	64/100:1/45/45	Red algae (*Gelidium elegans*)	547.3 ± 23.7 21.8 ± 11.1	52.1	[85]
	64/100:1/45/45	Red algae waste	315 ± 30.3/9 ± 3.1	13.76	[65]
	35/100:1/50/ 4140	Freeze- dried bacterial cellulose	599.60 ± 325.26/20.81 ± 7.66	-	[86]
	48/-/55/780	Tunicate cellulose	1073 ± 67/28 ± 46	-	[87]

Table 2. (Continued)

Acid used	Hydrolysis parameters	Cellulose source	CNCs size	Yield (wt%)	References
	Acid concentration/ acid:solid ratio/temperature/ time (wt%/v:m/°C/min)		Length/width (nm/nm)		
Hydrochloric acid	4 (M)/40:1/100/ 240	Tunicate cellulose	-	-	[88]
	6 (M)/60:1/110/ 180	MCC	262 ± 4/9 ± 3.1	93.7	[89]
	1 (M)/20:1/45/75	Curaua fiber	83 ± 24/7 ± 2	-	[90]
	3(M)/20:1/80/120	Oil palm empty fruit bunches	264 ± 43/12 ± 2	21	[91]
	4(M)/35:1/80/225	MCC	-/10 – 20	-	[92]
Hydrochloric acid (vapor)	35/-/-/720	Filter paper	216 ± 75/11 ± 5	54	[93]
Sulfuric acid + hydrochloric acid	98 wt% H_2SO_4 +37 Wt% HCl+ water (3:1:11 v/v)/-/80/ 420	Waste cotton cloth	28 – 470 /3 – 35	46.7	[40]
Acetic Acid + nitric acid	99% (v/v) acetic acid + 68% (v/v) HNO_3 (10:1 v/v)/25:1/60/120	Bamboo cellulose	6.5 – 20/5 – 7	-	[52]
Citric acid + hydrochloric acid	3M citric acid + 6M HCl (9:1 v/v)/50:1/80/120	MCC	249.8 ±22.4/15.6 ± 2.5	89.5	[94]
Phosphoric acid	70/130:1/100/80	Whatman filter paper	475/-	~40	[16]
	60-62% (v/v)/275:1/100/ 210	Tunicate cellulose	2300 ± 760/35 ± 12	-	[95]

Acid used	Hydrolysis parameters	Cellulose source	CNCs size	Yield (wt%)	References
	Acid concentration/ acid:solid ratio/temperature/ time (wt%/v:m/°C/min)		Length/width (nm/nm)		
Phosphoric acid	73.9/50:1/100/90	Whatman filter paper	506 ± 127/31 ± 14	-	[96]
Phosphotungstic Acid	75/80:1/90/1800	Bleached hardwood pulp	600 – 800/15 – 25	60.5	[97]
Formic Acid	88/30:1/90/720	Bleached eucalyptus karft pulp	141/11	~25	[98]
	98/50:1/95/240	Bleached chemical pulp	50 – 300 /2 – 4	70.21 ± 0.13	[78]
Nitric acid	6M/25:1/40/240	Filter paper	316 ± 40/21 ± 8	-	[99]
Oxalic acid	60/8:1/100/90	Bleached eucalyptus karft pulp	273/10	-	[100]
	50/10:1(m:m)/ 100/60	Bleached eucalyptus karft pulp	547/5	36.9	[101]
Hydrobromic acid	2.5(M)/50:1/80/180	Green algae (*Cladophora rupestris*)	483 ± 88/20 ± 4.4	70.2 ± 20	[66]
Maleic Acid	60/10:1/120/120	Unbleached hardwood pulp	232.9/20.7	5.94	[102]
Citric Acid	80/50:1/100/240/	Sugarcane bagasse pulp	251 ± 52/21 ± 6	32.1 ± 1.1	[103]

Satyamurthy et al. prepared cellulose nanowhiskers (CNWs) by hydrolyzing cotton fibers derived MCC with 5% inoculum of fungus *Trichoderma reesei* [108]. Although CNW yield was very negligible until two days of incubation of fungus, maximum CNW yield of 22% was achieved after 5 days of incubation of *T. reesei* at 30°C. The study also reported that the average dimensions of CNWs produced were length: 120.27±36.25 and width: 40.74 ±7.59 nm [108]. Similarly, Zhang et al. used microbial hydrolysis process using *T. reesei* to produce CNCs from bamboo fibers [109]. The authors reported that rod-like shaped nanocrystalline cellulose (NCC) with an average diameter of 24.7 nm and length of 286 nm were successfully prepared [109]. Anderson et al. sought to produce CNC with similar morphologies obtained from acid hydrolysis by enzymatic method [110]. They examined the production of CNCs by employing different enzymes with endo-glucanase activity from different feedstocks: fully bleached kraft pulp (FBK), thermomechanically processed pulps from live, and dead wood. The authors reported that the incubation of cellulase from *Aspergillus niger* with FBK at 24 h and longer produced CNCs. Unlike CNCs obtained from sulfuric acid hydrolysis, the CNCs produced exhibited poor dispersion in the aqueous medium, similar to the behavior of CNCs produced using hydrochloric acid. The study estimated the CNC yield of 10% from a 62 h incubation of *Aspergillus niger* cellulase with FBK at a ratio of 10:1 FBK to the cellulase [110]. In another study, Cui et al. produced a rod-like structured NCC from MCC with length of 40 – 50 nm and width of 5 – 8 nm at relatively high yield of 22.5% by employing enzymatic hydrolysis coupled with ultrasonic treatment [106]. The optimal conditions for maximum CNC yield were reaction temperature of 50°C, hydrolysis time of 120 h and 10 ultrasonic treatment each for 60 min. The CNC yield decreased to 17% in the absence of ultrasonic treatment [106].

The obvious advantage of synthesizing CNCs using enzymatic/microbial hydrolysis is that it obviates the use of harsh chemicals (corrosive acids) and thus the process is environmentally friendly. However, the enzymatic hydrolysis has disadvantage over the traditional process (acid hydrolysis) in terms of efficiency because the CNC yield is low (usually ~ 10%) and the process is very time consuming (e.g., tens of minutes in acid hydrolysis versus days in enzymatic hydrolysis). Moreover, pretreatments such as mechanical grinding are necessary in order to liberate cellulose into elementary fibrils to enhance the enzyme accessibility. However, some studies have shown that CNCs can be produced by enzymatic hydrolysis methods at yields comparable to the conventional sulfuric acid hydrolysis in short hydrolysis time. Filson et al. reported that cylindrical CNCs with diameter of 30 – 80 nm and length of 100 nm – 1.8 μm were produced when recycled pulp fibers were treated with endoglucanase (420 EGU/g fiber) at reaction temperature of 50°C for 45 min with either conventional or microwave heating. The CNC yield was 29 and 39% for conventional and microwave heating respectively [111]. Recently, Juárez-Luna and colleagues reported on the production of rod-shaped CNCs from water hyacinth stem cellulose by enzymatic hydrolysis using commercially available cellulase enzyme complex marketed by Novozymems in a relatively short period of time [112]. The researchers treated cellulose with cellulase at 50°C (0.12 U_{CM} enzyme/mg cellulose) for different time period. The study concluded that 120 min of treatment was optimum for the production of CNCs by hydrolyzing the amorphous domains without hampering the crystalline cellulose. The average diameter and length of the CNCs produced at optimal hydrolysis conditions were 22.5±6.9 and 120.4±64.4 nm respectively [112].

3.2.3. Oxidative Process

Owing to the high reactivity of hydroxyl groups present in cellulose, these groups can be oxidized to aldehydes, ketones, and carboxylic groups by employing suitable oxidizing agents. In the process, native cellulose fibrillar structure is destructured and the degree of polymerization is reduced. Several studies have exploited this oxidative chemistry for the preparation and modification of nanocellulose/cellulose nanocrystals (CNCs) [113]. CNCs produced through oxidative route introduced carboxylate groups in one step process and therefore they exhibit excellent dispersion in aqueous medium [113–115]. In addition, the carboxylate groups on CNCs facilitate the chemical modification and functionalization of CNCs. Depending on the oxidizing agents used, cellulose can be selectively oxidized either at C6 primary alcohol without altering the ring structure (e.g., 2,2,6,6-tetramethylpiperidine-1-oxyl (TEMPO)-mediated oxidation (simply TEMPO-oxidation)) or through cleavage of the glucopyranose ring at C2 – C3 position, which results in dialdehyde cellulose (e.g., periodate oxidation). The oxidation of cellulose using ammonium persulfate (APS), is nonselective [113].

Hirota et al. prepared cellulose II nanocrystals from mercerized wood pulp by TEMPO mediated oxidation [116]. The researchers used 4-acetamide-TEMPO/NaClO/$NaClO_2$ system at pH 4.8 and 60°C. Very well dispersible CNCs, with length and width in the range of 100-200 and 4-7 nm respectively, were obtained when the oxidized product was subjected to ultrasonication treatment for 5 min. This study also showed that the CNCs produced directly via TEMPO oxidation had higher carboxylate content than CNCs prepared by hydrolyzing with HCl followed by TEMPO oxidation [116]. Another study by Peyre et al. reported on the production of CNCs form MCC, albeit in very low yield (3.9% wt fraction of the oxidized product), using TEMPO-oxidation (TEMPO/ClO_2/NaClO) for 22 h at pH 8 and 25°C [114]. CNCs with lengths of 40-130 nm and widths of 5-9 nm could be separated in supernatant and the rest of the oxidized product contained much larger size particles ranging from few microns to 100 microns in length [114]. Chen and coworkers showed that CNCs, which were sterically stabilized by short chains of dialdehyde modified cellulose one on each ends, were produced along with fibrous cellulose and dialdehyde modified cellulose when kraft softwood pulp was subjected to periodate oxidation ($NaIO_4$/NaCl/water) for three different periods of time (26, 42, and 84 h at room temperature) [117]. While the CNCs had similar diameters (5-10 nm), irrespective of the oxidation period, the average length of the CNCs decreased from 590 nm for 26 h to 100 nm for the 84 h oxidation. The wt% (based on the oxidized products) of CNCs increased from 35.5 to 43.6% when the oxidation period was increased from 26 h to 84 h [117].

Recently, a common strong oxidizing agent, ammonium persulfate (APS) has gained popularity among researchers for the production of CNCs because of its low long-term toxicity, high water solubility and low cost compared to other oxidizing agents used for the production of CNCs [118–122]. Additionally, while highly purified cellulose source material is needed in TEMPO or periodate oxidation, cellulose source material purified to a lesser degree can be used with APS for CNCs preparation because peroxide and persulfate radicals generated during the oxidation process are simultaneously used in delignification and disintegration of cellulose fibrillar structure [118, 122].

Lueng et al. demonstrated that carboxylated CNCs with high morphological uniformity can be produced in one-step process from different cellulose sources when they were heated in 1M APS at 60°C for 3 h (bacterial cellulose) or 16 h (plant-derived substrates) [122]. The investigators also mentioned that CNCs could be produced from the substrates containing up to 20% lignin. The yield of the CNCs ranged from modest (14% for bacterial cellulose) to very high (81% for filter paper) depending on the substrates. The average diameter and length of the CNCs ranged from 3.8 ± 0.1 to 6.7 ± 0.3 and 88 ± 5 to 296 ± 16 respectively depending on the substrates [122]. In a recent study, Bashar et al. used jute fibers without scouring and bleaching pretreatments to produce CNCs from 1M APS oxidation at 60-65°C for five different time periods (from 6 h to 24 h) [118]. CNCs with "crystal-like" morphologies were produced in all cases and the average length of CNCs decreased as the oxidation time increased while the average CNCs diameter (5.2 nm) was not affected by the oxidation time. The average lengths of CNCs were 458 nm, 425 nm, 308 nm, 306 nm, and 291 nm for 6, 8, 12, 16, and 24 h respectively. The CNCs yield also followed similar trend and decreased from 64% to 54% when the oxidation time increased from 6 h to 24 h [118].

Despite all the advantages of APS oxidation method over other methods for the preparation of CNCs, it is still time-consuming process. In a typical method of the preparation of CNCs through APS oxidation, the reaction is carried for 16 h to 24 h [115, 119, 123, 124]. However, recently, Ye et al. showed that carboxylated spherical CNCs from waste viscose fibers, with a relatively high yield (37.9%) can be produced with hydrothermal APS oxidation of cellulose in 1M APS at 80°C in 4 h, which is a significantly short period of time compared to 16 h to 24 h time period required to obtain

CNCs in common APS oxidation procedure. The average diameter of the spherical CNCs obtained from optimum oxidation period of 4 h was 46 ± 9 nm [121]. In another study, Zhou et al. reported that carboxylated CNCs with high yield were obtained using potassium permanganate as oxidant at much lower dose (~0.33 M) and relatively short period of time [125]. The researchers carried out oxidation of cotton cellulose pulp in 1 wt% sulfuric acid medium at 50°C using $KMnO_4$ and oxalic acid as oxidizing and reducing agents in a mass ratio of 2:1 respectively. The CNCs yield increased from 48% after 4 h oxidation to maximum yield of 68% after 8 h oxidation. The average length and diameter of CNCs after 4 h oxidation were 299.8 ± 18.2 nm and 21.5 ± 3 nm, and the corresponding dimensions of CNCs after 8 h oxidation decreased to 210.3 ± 18.7 nm and 15.9 ± 2.9 nm [125].

Moreover, the usefulness of other common oxidants such as hydrogen peroxide (H_2O_2) for successful preparation of CNCs has been demonstrated. Koshani and colleagues produced carboxylated cellulose nanocrystals from the oxidation of softwood pulp using a reaction system consisting of ~10.9 M H_2O_2 as an oxidant and 0.34 μM copper(II) sulfate pentahydrate as a catalyst at 60°C for 72 h [126]. The reaction proceeded at acidic pH of 1-2 maintained by the addition of 1M HCl. The CNCs yield was 54%, which increased to 81% when the reaction was coupled with ultrasound treatment. The obtained CNCs had an average length and width of 263 and 23 nm respectively [126].

3.2.4. Other Methods

Apart from the above discussed existing dominant methods for the preparation of CNCs, other methods such as hydrolysis with ionic liquids, subcritical water, and purely mechanical means e.g., high-pressure homogenization have also been reported [127, 128]. The quest for new methods primarily stems from the necessity to make the process more environmentally friendly or "green" as large quantities of harsh chemicals such as corrosive acids and oxidants are released in the environment [128].

Ionic liquids are generally asymmetrical organic salts comprised of a bulky low charge density organic cation and a low charge density inorganic and organic anion. They have relatively low melting temperature (<100°C). Even though they have been well established as powerful and green cellulose solvents as well as reagents for the pretreatments of lignocellulosic biomass because of their higher biomass processing ability, chemical stability, low vapor pressure, and recyclability [129–131], a few studies have also shown that cellulose nanocrystals were directly obtained when cellulose was treated with certain ionic liquids. Tan et al. prepared CNCs from MCC using 1-butyl-3-methylimidazolium hydrogen sulfate ($BminHSO_4$) at 70 – 100°C for 90 min [132]. In a similar mechanistic fashion to acid hydrolysis as claimed in the study, glycosidic bonds of cellulose chains in the amorphous domains were preferentially attacked by HSO_4^- dissociated from the ionic liquid and CNCs were ultimately extracted when the reaction product was subjected to sonication. The CNCs produced from higher temperature (90 and 100°C) had lower diameter and refined fibrillar structure. The diameter and length of CNCs produced from optimum temperature of 90°C were 15 – 20 and 70 – 80 nm [132]. In another study, Abushamala and coworkers reported on the production of CNCs by treating dewaxed wood powder from Angelim vermelho (*Dinizia excelsa*) with a popular ionic liquid 1-ethyl-3-methylimidazolium acetate (EmimOAc) for two identical pulping cycles, each consisting of 2 h at 60°C [133]. The production of CNCs from wood powder, which originally contained 33% lignin and 19% hemicellulose, was attributed to the capability of EmimOAc to dissolve lignin but only to swell cellulose, acetylation of wood leading to its decreased intermolecular cohesion, and to catalyze the hydrolytic cleavage of the amorphous domains of cellulose simultaneously. The produced CNCs had an average length and height dimensions of 117 ± 30 and 1.9 ± 0.5 nm respectively. These CNCs amounted to 60 and 20% of the produced pulp and original wood mass respectively [133]. Since the recovery up to 95% has been reported for ionic liquids, these studies demonstrating the capability of ionic liquids for the direct production of CNCs hint to the potential use of ionic liquids for economically and environmentally friendly production of CNCs in the future.

Controlled hydrolysis of cellulose by water for the production of CNCs would be a very interesting strategy in terms of environmental friendliness, lowered amount of harmful effluents and potential cost reduction of the process. In fact, the ability of water to completely hydrolyze cellulose to sugars at elevated temperature and pressure (e.g., supercritical water) has been shown [134]. In an interesting novel study, Novo and colleagues achieved partial hydrolysis of commercial MCC (Avicel) to produce CNCs by treating it with subcritical water (120°C and 20.3 MPa) for 60 min. The CNCs produced possessed an average length and width of 242 ± 98 nm and 55 ± 20 nm. The CNC yield was 21.9% [127]. The same group investigated the effects of pressure and temperature of the subcritical water in the range of 8.1 – 20.3 MPa and 120 – 200°C keeping the duration of the reaction constant at 60 min as in their previous study. CNCs were obtained in all different reaction conditions investigated. However, the study revealed that the yield was directly proportional to the pressure employed while the temperature had directly and inversely proportional effect on the stability of the CNCs suspension and the whiteness of the CNCs respectively[135].

Although the use of mechanical means to disintegrate the cellulose native fibrillar structure for the production of enmeshed individual or bundle of fibrils (i.e., nanofibers which contain both the crystalline and amorphous domains) is commonplace, the viability of solely mechanical means for the production of CNCs was demonstrated by Park et al.in a recent study [128]. In their study, scoured and bleached cotton cellulose powder slurry was subjected to multiple cycles of high-pressure homogenization (HPH) at a rate of 100 mL/min under 1,400 bar at a certain temperature. CNCs with surprisingly high yield of 80% were produced with 20 passes at 80°C. The average length and width of CNCs were 177.5 ± 123.7 nm and 7.7 ± 3 nm respectively [128]. The process, by virtue of being simple and chemical free yet producing CNCs at high yield shows potential for an efficient and sustainable green route of CNCs production. Nonetheless, the energy requirement in the process should be taken into consideration and further studies to address the energy consumption issues are needed.

4. APPLICATIONS OF CNCS

4.1. Applications in Bioscaffold Materials

Scaffold materials should be capable of regenerating lost tissue. The design and performance of scaffold materials are adapted as per the need of target tissue. The pore size, pore distribution, interconnection are primarily taken into consideration to perform scaffold design. The unique properties of the scaffold depend on the selection of materials, which promote cellular responses [136]. The applications of conventional porous materials (silica and carbon-based materials) are limited due to their brittleness, toxicity, high conductivity, high manufacturing cost, etc. [137]. This created a demand for cellulosic nanomaterials for making porous biomaterials, such as, cellulosic scaffold, whose specific characteristics have made it promising in biomedical applications [138]. Having the properties of biodegradability, biocompatibility, and low-toxicity, CNCs have been widely investigated for use in the production of biomedical materials [139,140]. The use of nanocellulose in bioscaffolds can be in the forms of sponges, hydrogels, membranes, electrospun nanofibers, and composites [6, 141, 142]. The application of scaffold materials in tissue engineering is challenged by the sensitivity and the complex biological system of the human body [143]. It requires the biocompatibility of scaffold materials with good mechanical properties [144, 145]. Scaffold promotes the regeneration of new tissues. In particular, the use of scaffolds in bone engineering requires very good porosity, which directs cell penetration and support proper vascularization of the ingrown tissue [146–148]. Inclusion of CNC in the scaffolds is very promising for tissue engineering, particularly for culturing bone cell. CNC based scaffolds support better adhesion and proliferation of the cells and exhibit lower cytotoxicity [136, 149]. The presence of functional groups on the surface of cellulose nanoparticles works as a facile platform for its potential chemical modification of the surface. It allows tremendous possibility of using cellulose nanoparticles in different applications of scaffolds from zero dimension to 3-D materials [150]. CNC based 3-D scaffolds have gained considerable interests because of their suitability to mimic extracellular matrix, though there is a large number of polymeric materials for scaffold preparation [149]. CNC based porous scaffolds have been widely adapted in the applications of drug delivery systems, packaging, nanocomposites, particularly in tissue engineering as obvious structural components [151].

Shaheen et al. investigated the effect of inclusion of CNCs in the composites of 3-D scaffolds [152]. They prepared highly porous scaffolds by incorporating different concentrations of CNCs (0.5%, 1%, and 2%) in the composite of chitosan/ alginate/ hydroxyapatite through electrostatic gelation process. They found a remarkable enhancement in the properties of scaffolds containing 1% CNCs. The porosity of the scaffold composites with 1% CNCs increased by 93.6% with a swelling ratio of 110%. In addition, the cell viability of osteoblast cells cultured on that scaffold was almost 100% [152]. In a similar study, Li et al. designed a scaffold of CNC reinforced collagen-based composite, with different concentration of CNCs[153]. They found a good dispersion of CNCs in the matrix of collagen and did not observe any aggregation up to 7% of CNCs. The study reported 2-folds improvement in the swelling capacity of the composite film (with the loading of 7% CNCs) than the pure collagen film. Furthermore, the interaction between CNCs and collagen allowed transferring stress through their interface, which resulted in improvement of the tensile strength of the composite by 2.4 times. They also performed *in vitro* cell culture of 3T3 fibroblasts cell on the composite. The results showed the biocompatibility of the scaffold without any negative effect of cytotoxicity; rather it facilitated cell viability, cell adhesion, and proliferation [153]. In recent studies, Abraham et al. prepared acetylated cellulose nanocrystals (Ac-NC), followed by developing highly porous and ultralight cellulosic scaffolds through freeze-drying ice-templating of highly esterified (Degree of substitution = 2.18) cellulose nanocrystals (<1 wt%) [154, 155]. They investigated the functionality of the novel scaffolds. They reported multifunctional properties of the scaffolds, including hydrophobicity, oleophilicity, and lipophilicity. The scaffolds displayed the ability for selective absorption of hydrophobic proteins, milkfat, different organic solvents, and oils. Moreover, the thermal stability of the scaffolds from modified CNCs was 15.4% higher than the scaffolds with unmodified CNCs. The potential use of the multifunctional cellulosic scaffolds can be in diverse applications, such as scaffold for tissue engineering, drug delivery, accumulation of oil and fat [154, 155].. The use of CNC based scaffolds can also be found in biosensors or biocatalysts [156, 157]. They help enhance the activity of active compounds. A scaffold of CNC/PVA (cellulose nanocrystals/polyvinyl alcohol) composite functionalized with fluorescein was developed by Schyrr et al. [156]. Taking the benefits of CNCs' high aspect ratio, they prepared CNC/PVA composite with high surface area and porosity followed by surface modification with thiolated fluorescein-substituted lysine (FL-SH), which was sensitive to changes in pH. The abundant presence of hydroxyl groups on the surface led to a high degree of binding fluorophores, which allowed detecting pH change rapidly. Their findings suggest the possibility of developing new functional biosensor scaffolds based on CNCs [156]. In another study, He et al. successfully prepared a scaffold of CNC reinforced electrospun all-cellulose nanocomposite [142]. The incorporation of CNCs helped create a uniform morphology in the scaffold with an average fiber diameter ranging from 212 to 221 nm. They reported that with the loading of 20% CNCs in the composite, the interfacial bond between CNCs and the cellulose matrix was very strong, which enhanced the tensile strength and the elastic modulus by 101.7% and 171.6%, respectively. Furthermore, the incorporation of CNCs improved the thermal stability of the composite. In addition, the scaffold demonstrated its biocompatibility to human dental follicle cells (hDFCs). These unique properties of the scaffold make it potentially suitable for artificial blood vessels, which need to withstand high shear stress under blood flowing [142].

4.2. Applications as Reinforcement Fillers

Current demand for greener products is driving the research interest in renewable and non-petroleum based products. Renewable cellulose has gained special attention for its nanostructure, CNC. CNCs, as reinforcing fillers, are significantly potential in the field of nanocomposite materials due to their unique properties, such as high surface area, reactive surfaces, high aspect ratio, low density, high specific strength, etc. [158–160]. Even at lower volume fractions, their reinforcing effect can be very significant to improve the mechanical properties of the resulting nanocomposites [161, 162]. CNCs have been widely used for reinforcing a large number of polymers and biopolymer matrices, such as polyethylene, poly(caprolactone), polyvinyl chloride, polyurethane, thermoplastic starch, soy protein, and poly lactic acid [163–168]. The limitations in the mechanical performance of the polymeric materials can be improved by reinforcing the polymer matrix with CNCs.

The reinforcing effect of CNCs was first reported by Favier et al. [161], followed by a great deal of research in the field of CNC-reinforced bio nanocomposite applications. Lu et al. investigated the effect of reinforcing thermoplastic-starch film with CNCs on its mechanical performance [168]. They prepared nanocomposites of plasticized starch (PS) with different concentrations of CNCs. They reported that the synergistic interactions between PS matrix and CNC particles significantly improved the strength of the composites. They reported an increment in the tensile strength from 2.8 MPa to 6.9 MPa and an increment in the Young modulus from 56 MPa to 480 MPa upon addition of 40% CNCs in the PS matrix [168]. In a similar study, Zoppe et al. prepared a composite of PCL (polycaprolactone)/CNC nanofibers by a chemical grafting of hydrophilic CNCs on electrospun PCL [169]. The chemical grafting led to a better dispersion of CNCs in the matrix of PCL, therefore the observed morphology of the resulting composite was uniform. They reported that the composite reinforced with 2.5% CNCs showed 1.5 folds increase in Young modulus [169]. Paralikar et al. prepared membranes of poly(vinyl alcohol) (PVOH)/CNCs composite and investigated the mechanical performance and the barrier properties of the membranes [170]. They cross-linked the membranes with poly(acrylic acid) (PAA), followed by a heat treatment at 170°C to improve the water barrier properties. They reported that the highest mechanical performance of the membrane was obtained with 10% CNCs/10% PAA/80% PVOH, while the tensile strength was 150% higher than pure PVOH. The tensile modulus increased with the increase of CNC concentration up to 10% and the modulus decreased above 10% loading of CNCs. The decrease in tensile modulus at higher CNCs loading concentration could be due to CNCs agglomeration. Furthermore, the addition of CNCs improved the barrier properties to the flow of moisture. The incorporation of 10% CNCs decreased the rate of water vapor transmission to less than 50%, which was even better than water resistance exhibited by the incorporation of 10% PAA crosslinker [170]. Their findings demonstrated the possibility of developing membrane with improved barrier properties. Recently, the use of CNCs, as strength enhancer, to reinforce cementitious materials has gained special attention [171, 172]. Cao et al.reported that the reinforcement with only 0.2% CNCs (with respect to the volume of cement) increased the flexural strength by ~30%, which was attributed to the alteration of the hydration reaction by CNCs. The addition of CNC particles led to uniform distribution and steric stabilization of cement particles in the mixture, for which cement particles could efficiently react with water [172]. From the findings of different studies, it is demonstrated that CNCs have a significant potential to tailor different properties of the CNC reinforced composites, such as mechanical, chemical, thermomechanical, barrier properties (air permeability and water vapor transmission rate) [62, 159, 170, 172, 173]. Therefore, the functionalities of CNC reinforced nanocomposites with improved tensile strength, elastic modulus, flexural strength, impact resistance, barrier properties, etc. allow them to be used in highly functional applications, including ultrathin films, filtration membranes, paper-based products, textiles, supercapacitors, packaging materials, food packaging [151, 173–176].

4.3. Applications as Rheology Modifier

The processability of polymers depends on their rheological properties [177]. Rheology modifiers help achieve certain characteristics for a particular application. CNCs, as an effective rheology modifier, have been reported to be used in different applications [62, 177]. The influence of

CNCs as rheology modifier on the viscosity and the viscoelastic (loss modulus and storage modulus) behavior of nanocomposites has been reported in different studies [62, 163, 172, 177–182]. Marcovich et al. reported that the incorporation of CNCs (5 wt%) into the liquid mixtures of polyols increased the viscosity of the mixtures by 600 and 20 folds at low and high frequency respectively, as compared to that of neat polymer sample [163]. In a similar study by Cao et al., the effect of CNCs on the rheological behavior of cement paste was investigated [172]. They found a decrease in the yield stress of CNC-reinforced cement paste from 48.5 Pa to 15.9 Pa at a low concentration of CNC (0.04%), which could be due to the effect of steric stabilization. At higher concentration of CNCs (1.5%), the stress value increased to 600 Pa. This change in yield stress could be due to the effect of CNCs agglomeration [172]. CNCs have also the properties of shear thinning and therefore, theywork as natural additive to drilling fluids. CNCs cause a stronger shear thinning effect and provide additional functionality to drilling fluids. A study by Li et al. found a strong thinning effect of CNCs [181]. From their findings, they demonstrated the use of a small amount of CNCs in water-based drilling fluids/bentonite (WDF-BT) in order to reduce the amount of bentonite used in the fluid. They reported that the addition of CNCs to bentonite/water based drilling fluids created strong surface interactions, which allowed to form a dense core-shell structure around bentonite particles. It subsequently resulted in enhanced rheological properties (e.g., fluid loss, shear thinning) of the drilling fluids. The addition of only 1% CNCs to WDF-BT significantly reduced the loss of fluid volume by 44.2% [181]. A dramatic thickening effect was observed by Oguzlu et al. due to the incorporation of CNCs in carboxymethyl cellulose (CMC) solutions even at a lower concentration of CNCs [180]. The addition of CNC particles in the range of 0.33-2.02 m^3/m^3 concentration caused an increase in the viscosity of the CMC solution. The thickening behavior of CNCs can help operational processability of the system and tailor the performance of the product [180]. The functionalities of CNCs, as rheology modifier, open the possibility of using them in a wide range of material applications, such as cement, concrete, tiles, mortar, cosmetic products, pharmaceutical products, drilling fluids, coating formulations, different daily use products, etc. [172, 177, 180, 181].

4.4. Applications in Electrical and Electronics

CNCs hold great potential for green electronics. The combination of excellent mechanical properties such as low density, high tensile strength, high elastic modulus, and physical properties such as low surface roughness, good transparency and low coefficient of thermal expansion make them well suited for energy storage and photovoltaic applications. Therefore, the study on the CNC derived electronic devices for such applications has been energetically pursued over the last few years. Xuan et al. demonstrated that CNC aerogels can be used as universal substrates for the development of flexible energy storage devices with excellent capacitance retention [183]. They reported that in situ incorporation of polypyrrole nanofibers (PPy-NF), polypyrrole-coated carbon nanotubes (PPy-CNT), and manganese dioxide nanoparticles (MnO_2-NP) into the CNC aerogels give 3D supercapacitor devices with a low internal resistance, fast charge-discharge rates, and a good cycle stability. They found 84.19%, 61.66%, and 92.28% capacitance retention after 2000 cycles for PPy-NF, PPy-CNT, and MnO_2 -NP cells, respectively. Moreover, the hybrid CNC aerogels retained the capacitive behavior at scan rates up to 1000 mVs^{-1} in contrast to the scan rate of 500mVs^{-1} for other hybrid nanocellulose-based supercapacitors reported previously [183]. Guoyin and coworkers prepared and investigated the electrical properties of reduced graphene oxide (rGO)/CNC hybrid fibers [184]. They reported an increment in the specific capacitance from 5 F/g to 121.2 F/g upon the addition of 20% CNC on rGO. Furthermore, the hybrid fibers had a good life cycle with 92% capacitance retention after 1000 charge-discharge cycles [184]. Similarly, Bokhari et al. reported a specific capacitance of 397.7 F g^{-1} at the current density of 0.2 Ag^{-1} and 96.6% capacitance retention after 2000 cycles for the rGO-CNC-MnO_2 composite electrode material [185]. Flexible fiber shaped CNC-based supercapacitors with rGO and manganosite (MnO) nanoparticles were prepared by Yuan et al. [186] and it was reported that it had a capacitance of 1.59 Fcm^{-3} at the voltage window of 0.8 V and an energy density of 0.14 $mWhcm^{-3}$ at the power density of 4 mW cm^{-3}. In addition, they reported an excellent cycling stability (82% after 6000 cycles) and bending robustness of the assembled device [186]. The development of organic electronic devices such as organic solar cells is another important approach for the utilization of CNCs for electronic purposes. Zhou et al. fabricated organic solar cells using CNC substrate. They reported that the prepared solar cell had power conversion efficiency (PCE) of 2.7%, good rectification ratio in the dark and good recyclability [187]. They further increased the PCE of the solar cell to 3.8% by changing the electrodes while continuing the use of CNC substrate [188]. Similarly, Yuwawech et al. reported that the encapsulation of dye-sensitized solar cells with polyurethane (PU)/esterified CNCs lead to an extension of lifetime of the solar cells by more than 336 h without compromising the PCE [189]. Recently, CNCs have found their application in the development of sensors and light-emitting diodes (LEDs) as well. Sadasivuni et al. developed a CNC/GO-based proximity sensor with potential applications in non-touch screens for electronic appliances [190]. They found that the sensor can detect a human finger within 6 mm distance with good response and recovery time interval. Najafabadi et al. used CNCs to prepare an organic LED (OLED) [191]. The OLED fabricated on CNC substrate exhibit maximum efficacy of 53.7 cd/A at a luminance of 100 cd/m^2 and 41.7 cd/A at 1000 cd/m^2 with peak luminance of 74591 cd/m^2 [191]. This shows that CNCs can be used in a wide range of electrical applications giving a low cost, green, sustainable and environmentally-friendly alternatives for current sources of materials in different electrical and electronic applications.

CONCLUSION

Cellulose nanocrystals (CNCs) are highly crystalline rod-like nanoparticles of cellulose extracted after preferential removal of amorphous domains of cellulose microfibrils. Because of their special morphology, impressive mechanical properties, liquid crystalline behavior, and other useful characteristics, CNCs have been widely studied and investigated for various important applications such as scaffolds, reinforcement, polymer processing, electrical, and electronics. In this chapter, we extensively reviewed and discussed the sources of CNCs, different methods of their preparation and their applications in polymeric scaffolds, reinforcement, rheology modification of solutions, and electrical and electronics.

REFERENCES

[1] Habibi, Y., Lucia, L. A. and Rojas, O. J. (2010). Cellulose nanocrystals: chemistry, self-assembly, and applications. *Chemical Reviews,* 110: 3479-500.

[2] French, A. D. (2017). Glucose, not cellobiose, is the repeating unit of cellulose and why that is important. *Cellulose*, 24: 4605-4609.

[3] Moon, R. J., Martini, A., Nairn, J., Simonsen, J. and Youngblood, J. (2011). Cellulose nanomaterials review: structure, properties and nanocomposites. *Chemical Society Reviews*, 40: 3941-3994.

[4] Gibson, L. J. (2012). The hierarchical structure and mechanics of plant materials. *Journal of the Royal Society Interface*, 9: 2749-2766.

[5] Payne, C. M., Knott, B. C., Mayes, H. B., Hansson, H., Himmel, M. E., Sandgren, M., Ståhlberg, J. and Beckham, G. T. (2015). Fungal cellulases. *Chemical Reviews*, 115: 1308-1448.

[6] Lin, N. and Dufresne, A. (2014). Nanocellulose in biomedicine: Current status and future prospect. *European Polymer Journal*, 59: 302-325.

[7] Eichhorn, S. J., Rahatekar, S. S., Vignolini, S. and Windle, A. H. (2018). New horizons for cellulose nanotechnology. *Philosophical Transactions of the Royal Society A: Mathematical, Physical and Engineering Sciences*, 376: 20170200.

[8] Koshani, R. and Madadlou, A. (2017). A viewpoint on the gastrointestinal fate of cellulose nanocrystals. *Trends in Food Science & Technology*, 71: 268-273.

[9] Dufresne, A. (2013). Nanocellulose: a new ageless bionanomaterial.

Materials Today, 16: 220-227.

[10] Dufresne, A. (2010). Processing of polymer nanocomposites reinforced with polysaccharide nanocrystals. *Molecules*, 15: 4111-4128.

[11] El Achaby, M., Kassab, Z., Barakat, A. and Aboulkas, A. (2018). Alfa fibers as viable sustainable source for cellulose nanocrystals extraction: Application for improving the tensile properties of biopolymer nanocomposite films. *Industrial Crops and Products*, 112: 499-510.

[12] Klemm, D., Kramer, F., Moritz, S., Lindström, T., Ankerfors, M., Gray, D. and Dorris, A. (2011). Nanocelluloses: a new family of nature-based materials. *Angewandte Chemie - International Edition*, 50: 5438-5466.

[13] Sacui, I. A., Nieuwendaal, R. C., Burnett, D. J., Stranick, S. J., Jorfi, M., Weder, C., Foster, E. J., Olsson, R. T. and Gilman, J. W. (2014). Comparison of the properties of cellulose nanocrystals and cellulose nanofibrils isolated from bacteria, tunicate, and wood processed using acid, enzymatic, mechanical, and oxidative methods. *ACS Applied Materials & Interfaces*, 6: 6127-6138.

[14] Beck-Candanedo, S., Roman, M. and Gray, D. G. (2005). Effect of reaction conditions on the properties and behavior of wood cellulose nanocrystal suspensions. *Biomacromolecules*, 6: 1048-1054.

[15] Trache, D., Hussin, M. H., Haafiz, M. K. M. and Thakur, V. K. (2017). Recent progress in cellulose nanocrystals: sources and production. *Nanoscale*, 9: 1763-1786.

[16] Vanderfleet, O. M., Osorio, D. A. and Cranston, E. D. (2017). Optimization of Cellulose Nanocrystal Length and Surface Charge Density through Phosphoric Acid Hydrolysis. *Philisophical Transactions of the Royal Society A: Mathematical, Physical and Engineering Sciences*, 376: 20170041.

[17] Qi, H. (2017). *Novel Functional Materials Based on Cellulose*, Springer.

[18] Murr, L. E. (2015). Examples of Natural Composites and Composite Structures. In: *Handbook of Materials Structures, Properties,*

Processing and Performance, Springer International Publishing: 425-449.

[19] Morais, J. P. S., de Freitas Rosa, M., Nascimento, L. D., do Nascimento, D. M. and Cassales, A. R. (2013). Extraction and characterization of nanocellulose structures from raw cotton linter. *Carbohydrate Polymers*, 91: 229-235.

[20] Hu, Y. and Abidi, N. (2016). Distinct chiral nematic self-assembling behavior caused by different size-unified cellulose nanocrystals via a multistage separation. *Langmuir*, 32: 9863-9872.

[21] Mtibe, A., Mandlevu, Y., Linganiso, L. Z. and Anandjiwala, R. D. (2015). Extraction of cellulose nanowhiskers from flax fibres and their reinforcing effect on poly(furfuryl) alcohol. *Journal of Biobased Materials and Bioenergy*, 9: 309-317.

[22] Luzi, F., Fortunati, E., Puglia, D., Lavorgna, M., Santulli, C., Kenny, J. M. and Torre, L. (2014). Optimized extraction of cellulose nanocrystals from pristine and carded hemp fibres. *Industrial Crops and Products*, 56 : 175-186.

[23] Mandal, A. and Chakrabarty, D. (2011). Isolation of nanocellulose from waste sugarcane bagasse (SCB) and its characterization. *Carbohydrate Polymers*, 86: 1291-1299.

[24] Mandal, A. and Chakrabarty, D. (2014). Studies on the mechanical, thermal, morphological and barrier properties of nanocomposites based on poly(vinyl alcohol) and nanocellulose from sugarcane bagasse. *Journal of Industrial and Engineering Chemistry*, 20: 462-473.

[25] Leão, R. M., Miléo, P. C., Maia, J. M. L. L. and Luz, S. M. (2017). Environmental and technical feasibility of cellulose nanocrystal manufacturing from sugarcane bagasse. *Carbohydrate Polymers*, 175: 518-529.

[26] Pasquini, D., de Morais Teixeira, E., da Silva Curvelo, A. A., Belgacem, M. N. and Dufresne, A. (2010). Extraction of cellulose whiskers from cassava bagasse and their applications as reinforcing agent in natural rubber. *Industrial Crops and Products*, 32: 486-490.

[27] Shaheen, T. I. and Emam, H. E. (2018). Sono-chemical synthesis of

cellulose nanocrystals from wood sawdust using acid hydrolysis. *Interantional Journal of Biological Macromolecules*, 107: 1599-1606.

[28] Rehman, N., de Miranda, M. I. G., Rosa, S. M. L., Pimentel, D. M., Nachtigall, S. M. B. and Bica, C. I. D. (2014). Cellulose and nanocellulose from maize straw: an insight on the crystal properties. *Journal of Polymers and the Environent*, 22: 252-259.

[29] Mueller, S., Weder, C. and Foster, E. J. (2014). Isolation of cellulose nanocrystals from pseudostems of banana plants. *RSC Advances*, 4: 907-915.

[30] Smyth, M., García, A., Rader, C., Foster, E. J. and Bras, J. (2017). Extraction and process analysis of high aspect ratio cellulose nanocrystals from corn (*Zea mays*) agricultural residue. *Industrial Crops and Products*, 108: 257-266.

[31] Liu, C., Li, B., Du, H., Lv, D., Zhang, Y., Yu, G., Mu, X. and Peng, H. (2016). Properties of nanocellulose isolated from corncob residue using sulfuric acid, formic acid, oxidative and mechanical methods. *Carbohydrate Polymers*, 151: 716-724.

[32] Rosa, M. F., Medeiros, E. S., Malmonge, J. A., Gregorski, K. S., Wood, D. F., Mattoso, L. H. C., Glenn, G., Orts, W. J. and Imam, S. H. (2010). Cellulose nanowhiskers from coconut husk fibers: Effect of preparation conditions on their thermal and morphological behavior. *Carbohydrate Polymers*, 81: 83-92.

[33] Barana, D., Salanti, A., Orlandi, M., Ali, D. S. and Zoia, L. (2016). Biorefinery process for the simultaneous recovery of lignin, hemicelluloses, cellulose nanocrystals and silica from rice husk and *Arundo donax*. *Industiral Crops and Products*, 86: 31-39.

[34] Johar, N. and Ahmad, I. (2012). Morphological, thermal, and mechanical properties of starch biocomposite films reinforced by cellulose nanocrystals from rice husks. *BioResources*, 7: 5469-5477.

[35] Johar, N., Ahmad, I. and Dufresne, A. (2012). Extraction, preparation and characterization of cellulose fibres and nanocrystals from rice husk. *Industrial Crops and Products*, 37: 93-99.

[36] Naduparambath, S., T.V., J., V., S., M.P., S., Balan, A. K. and E., P.

(2018). Isolation and characterisation of cellulose nanocrystals from sago seed shells. *Carbohydrate Polymers*, 180: 13-20.

[37] Kallel, F., Bettaieb, F., Khiari, R., García, A., Bras, J. and Chaabouni, S. E. (2016). Isolation and structural characterization of cellulose nanocrystals extracted from garlic straw residues. *Industrial Crops and Products*, 87: 287-296.

[38] Rhim, J. W., Reddy, J. P. and Luo, X. (2015). Isolation of cellulose nanocrystals from onion skin and their utilization for the preparation of agar-based bio-nanocomposites films. *Cellulose*, 22: 407-420.

[39] Oun, A. A. and Rhim, J. W. (2016). Isolation of cellulose nanocrystals from grain straws and their use for the preparation of carboxymethyl cellulose-based nanocomposite films. *Carbohydrate. Polymers*, 150: 187-200.

[40] Wang, Z., Yao, Z. J., Zhou, J. and Zhang, Y. (2017). Reuse of waste cotton cloth for the extraction of cellulose nanocrystals. *Carbohydrate Polymers*, 157: 945-952.

[41] Purkait, B. S., Ray, D., Sengupta, S., Kar, T., Mohanty, A. and Misra, M. (2011). Isolation of cellulose nanoparticles from sesame husk. *Industrial & Engineering Chemistry Research*, 50: 871-876.

[42] Wijaya, C. J., Saputra, S. N., Soetaredjo, F. E., Putro, J. N., Lin, C. X., Kurniawan, A., Ju, Y. H. and Ismadji, S. (2017). Cellulose nanocrystals from passion fruit peels waste as antibiotic drug carrier. *Carbohydrate Polymers*, 175: 370-376.

[43] Lu, P., and Hsieh, Y. L. (2012). Cellulose isolation and core-shell nanostructures of cellulose nanocrystals from chardonnay grape skins. *Carbohydrate Polymers*, 87: 2546-2553.

[44] Jiang, F. and Hsieh,Y. L. (2015). Cellulose nanocrystal isolation from tomato peels and assembled nanofibers. *Carbohydrate Polymers*, 122: 60-68.

[45] Chen, D., Lawton, D., Thomson, M. R. and Liu, Q. (2012). Biocomposites reinforced with cellulose nanocrystals derived from potato peel waste. *Carbohydrate Polymers*, 90: 709-716.

[46] Zain, N. F. M., Yusop, S. M. and Ahmad, I. (2015). Preparation and characterization of cellulose and nanocellulose from pomelo (*Citrus*

grandis) albedo. *Journal of Nutrition & Food Sciences*, 5: 10-13.

[47] dos Santos, R. M., Neto, W. P. F., Silvério, H. A., Martins, D. F., Dantas, N. O. and Pasquini, D. (2013). Cellulose nanocrystals from pineapple leaf, a new approach for the reuse of this agro-waste. *Industrial Crops and Products*, 50: 707-714.

[48] Henrique, M. A., Silvério, H. A., Neto, W. P. F. and Pasquini, D. (2013). Valorization of an agro-industrial waste, mango seed, by the extraction and characterization of its cellulose nanocrystals. *Journal of Environmental Management*, 121: 202-209.

[49] Marett, J., Aning, A. and Foster, E. J. (2017). The isolation of cellulose nanocrystals from pistachio shells via acid hydrolysis. *Industrial Crops and Products*, 109: 869-874.

[50] Bano, S. and Negi, Y. S. (2017). Studies on cellulose nanocrystals isolated from groundnut shells. *Carbohydrate Polymers*, 157, 1041-1049.

[51] Lu, Q., Lin, W., Tang, L., Wang, S., Chen, X. and Huang, B. A. (2015). A mechanochemical approach to manufacturing bamboo cellulose nanocrystals. *Journal of Materials Science*, 50: 611-619.

[52] Zhang, P. P., Tong, D. S., Lin, C. X., Yang, H. M., Zhong, Z. K., Yu, W. H., Wang, H. and Zhou, C. H. Effects of acid treatments on bamboo cellulose nanocrystals. *Asia-Pacific Journal of Chemical Engineering*, 9: 686-695.

[53] Cudjoe, E., Hunsen, M., Xue, Z., Way, A. E., Barrios, E., Olson, R. A., Hore, M. J. A. and Rowan, S. J. (2017). Miscanthus Giganteus: a commercially viable sustainable source of cellulose nanocrystals. *Carbohydrate Polymers*, 155: 230-241.

[54] Robles, E., Fernández-Rodríguez, J., Barbosa, A. M., Gordobil, O., Carreño, N. L. V. and Labidi, J. (2018). Production of cellulose nanoparticles from blue agave waste treated with environmentally friendly processes. *Carbohydrate Polymers*, 183: 294-302.

[55] Sheltami, R. M., Abdullah, I., Ahmad, I., Dufresne, A. and Kargarzadeh, H. (2012). Extraction of cellulose nanocrystals from mengkuang leaves (*Pandanus tectorius*). *Carbohydrate Polymers*, 88: 772-779.

[56] Li, R., Fei, J., Cai, Y., Li, Y., Feng, J. and Yao, J. (2009). Cellulose whiskers extracted from mulberry: A novel biomass production. *Carbohydrate Polymers*, 76: 94-99.

[57] Nagalakshmaiah, M., El Kissi, N., Mortha, G. and Dufrense, A. (2016). Structural investigation of cellulose nanocrystals extracted from chili leftover and their reinforcement in cariflex-IR rubber latex. *Carbohydrate Polymers*, 136: 945-954.

[58] Malladi, R., Nagalakshmaiah, M., Robert, M. and Elkoun, S. (2018). Importance of agriculture and industrial waste in the field of nano cellulose and its recent industrial developments: a review. *ACS Sustainable Chemistry & Engineering*, 6: 2807-2828.

[59] Saichana, N., Matsushita, K., Adachi, O., Frébort, I. and Frebortova, J. (2015) Acetic acid bacteria: a group of bacteria with versatile biotechnological applications. *Biotechnology Advances*, 6: 1260-1271.

[60] Reiniati, I., Hrymak, A. N. and Margaritis, A. (2017). Recent developments in the production and applications of bacterial cellulose fibers and nanocrystals. *Critical Reviews in Biotechnology*, 37: 510-524.

[61] Singhsa, P., Narain, R. and Manuspiya, H. (2018). Bacterial cellulose nanocrystals (BCNC) preparation and characterization from three bacterial cellulose sources and development of functionalized BCNCs as nucleic acid delivery systems," *ACS Applied Nano Materials*, 1: 209-221.

[62] Moon, R. J., Schueneman, G. T. and Simonsen, J. (2016). Overview of cellulose nanomaterials, their capabilities and applications. *The Journal of the Minerals, Metals and Materials Society*, 68: 2383-2394.

[63] Gatenholm, P. and Klemm, D. (2010). Bacterial nanocellulose as a renewable material for biomedical applications. *MRS Bulletin*, 35: 208-213.

[64] Koyama, M. A. K. I. K. O., Sugiyama, J. U. N. J. I. and Itoh, T. A. K. A. O. (1997). Systematic survey on crystalline features of algal celluloses. *Cellulose*, 4: 147-160.

[65] El Achaby, M., Kassab, Z., Aboulkas, A., Gaillard, C. and Barakat, A. (2018). Reuse of red algae waste for the production of cellulose nanocrystals and its application in polymer nanocomposites. *International Journal of Biological Macromolecules*, 106:681-691.

[66] Sucaldito, M. R. and Camacho, D. H. (2017). Characteristics of unique HBr-hydrolyzed cellulose nanocrystals from freshwater green algae (*Cladophora rupestris*) and its reinforcement in starch-based film. *Carbohydrate Polymers*, 169: 315-323.

[67] Zhang, T., Cheng, Q., Ye, D. and Chang, C. (2017). Tunicate cellulose nanocrystals reinforced nanocomposite hydrogels comprised by hybrid cross-linked networks. *Carbohydrate Polymers*, 169: 139-148.

[68] Khandelwal, M. and Windle, A. H. (2013). Self-assembly of bacterial and tunicate cellulose nanowhiskers. *Polymer*, 54: 5199-5206.

[69] Chen, W., Yu, H., Liu, Y., Hai, Y., Zhang, M. and Chen, P. (2011). Isolation and characterization of cellulose nanofibers from four plant cellulose fibers using a chemical-ultrasonic process. *Cellulose*, 18: 433-442.

[70] Luzi, F., Puglia, D., Sarasini, F., Tirillò, J., Maffei, G., Zuorro, A., Lavecchia, R., Kenny, J. M. and Torre, L. (2019). Valorization and extraction of cellulose nanocrystals from North African grass: *Ampelodesmos mauritanicus* (Diss). *Carbohydrate Polymers*, 209: 328-337.

[71] Ng, H. M., Sin, L. T., Tee, T. T., Bee, S. T., Hui, D., Low, C. Y. and Rahmat, A. R. (2015). Extraction of cellulose nanocrystals from plant sources for application as reinforcing agent in polymers. *Composites Part B: Engineering*, 75: 176-200.

[72] Pirich, C. L., Picheth, G. F., Machado, J. P. E., Sakakibara, C. N., Martin, A. A., de Freitas, R. A. and Sierakowski, M. R. (2019). Influence of mechanical pretreatment to isolate cellulose nanocrystals by sulfuric acid hydrolysis. *International journal of biological macromolecules*, 130: 622-626.

[73] Kargarzadeh, H., Ioelovich, M., Ahmad, I., Thomas, S. and Dufresne, A. (2017). Methods for extraction of nanocellulose from various

sources. *Handbook of Nanocellulose and Cellulose Nanocomposites*, 1-49.

[74] Shin, H. K., Jeun, J. P., Kim, H. B. and Kang, P. H. (2012). Isolation of cellulose fibers from kenaf using electron beam. *Radiation Physics and Chemistry*, 81: 936-940.

[75] Shi, J., Shi, S. Q., Barnes, H. M. and Pittman Jr, C. U. (2011). A chemical process for preparing cellulosic fibers hierarchically from kenaf bast fibers. *BioResources*, 6: 879-890.

[76] Postek, M. T., Vladár, A., Dagata, J., Farkas, N., Ming, B., Wagner, R., ... and Beecher, J. (2010). Development of the metrology and imaging of cellulose nanocrystals. *Measurement Science and Technology*, 22(2): 024005.

[77] Dufresne, A. (2017). *Nanocellulose: From Nature to High Performance Tailored Materials*. Walter de Gruyter GmbH & Co KG.

[78] Li, B., Xu, W., Kronlund, D., Määttänen, A., Liu, J., Smått, J. H., Peltonen, J. H., Willför, S., Mu, X. and Xu, C. (2015). Cellulose nanocrystals prepared via formic acid hydrolysis followed by TEMPO-mediated oxidation. *Carbohydrate Polymers*, 133: 605-612.

[79] Chen, L., Wang, Q., Hirth, K., Baez, C., Agarwal, U. P. and Zhu, J. Y. (2015). Tailoring the yield and characteristics of wood cellulose nanocrystals (CNC) using concentrated acid hydrolysis. *Cellulose*, 22: 1753-1762.

[80] Shamskar, K. R., Heidari, H. and Rashidi, A. (2016). Preparation and evaluation of nanocrystalline cellulose aerogels from raw cotton and cotton stalk. *Industrial Crops and Products*, 93: 203-211.

[81] Zhang, H., Qian, Y., Chen, S. and Zhao, Y. (2019). Physicochemical characteristics and emulsification properties of cellulose nanocrystals stabilized O/W pickering emulsions with high -OSO_3^- groups. *Food Hydrocolloides*, 96: 267–277.

[82] Jordan, J. H., Easson, M. W., Dien, B., Thompson, S. and Condon, B. D. (2019). Extraction and characterization of nanocellulose crystals from cotton gin motes and cotton gin waste. *Cellulose*, 26: 5959-5979.

[83] Wang, Q. Q., Zhu, J. Y., Reiner, R. S., Verrill, S. P., Baxa, U. and McNeil, S. E. (2012). Approaching zero cellulose loss in cellulose nanocrystal (CNC) production: recovery and characterization of cellulosic solid residues (CSR) and CNC. *Cellulose*, 19: 2033-2047.

[84] Zhao, Y., Zhang, Y., Lindström, M. E. and Li, J. (2015). Tunicate cellulose nanocrystals: preparation, neat films and nanocomposite films with glucomannans. *Carbohydrate Polymers*, 117: 286-296.

[85] Chen, Y. W., Lee, H. V., Juan, J. C. and Phang, S. M. (2016). Production of new cellulose nanomaterial from red algae marine biomass *Gelidium elegans*. *Carbohydrate Polymers*, 151: 1210-1219.

[86] Martínez-Sanz, M., Lopez-Rubio, A. and Lagaron, J. M. (2011). Optimization of the nanofabrication by acid hydrolysis of bacterial cellulose nanowhiskers. *Carbohydrate Polymers*, 85: 228-236.

[87] Elazzouzi-Hafraoui, S., Nishiyama, Y., Putaux, J. L., Heux, L., Dubreuil, F. and Rochas, C. (2008). The shape and size distribution of crystalline nanoparticles prepared by acid hydrolysis of native cellulose. *Biomacromolecules*, 9: 57-65.

[88] Zhang, T., Zuo, T., Hu, D. and Chang, C. (2017). Dual physically cross-linked nanocomposite hydrogels reinforced by tunicate cellulose nanocrystals with high toughness and good self-recoverability. *ACS Applied Materials & Interfaces*, 9: 24230-24237.

[89] Yu, H., Qin, Z., Liang, B., Liu, N., Zhou, Z. and Chen, L. (2013). Facile extraction of thermally stable cellulose nanocrystals with a high yield of 93% through hydrochloric acid hydrolysis under hydrothermal conditions. *Journal of Materials Chemistry A*, 1: 3938-3944.

[90] Corradini, E., Pineda, E. A. G., Correa, A. C., Teixeira, E. M. and Mattoso, L. H. C. (2016). Thermal stability of cellulose nanocrystals from curaua fiber isolated by acid hydrolysis. *Cellulose Chemistry and Technology*, 50: 737-743.

[91] Hastuti, N., Kanomata, K. and Kitaoka, T. (2018). Hydrochloric Acid Hydrolysis of Pulps from Oil Palm Empty Fruit Bunches to Produce Cellulose Nanocrystals. *Journal of Polymers and the Environment*, 26: 3698-3709.

[92] Naseri, N., Mathew, A. P., Girandon, L., Fröhlich, M. and Oksman, K. (2015). Porous electrospun nanocomposite mats based on chitosan–cellulose nanocrystals for wound dressing: effect of surface characteristics of nanocrystals. *Cellulose*, 22: 521-534.

[93] Lorenz, M., Sattler, S., Reza, M., Bismarck, A. and Kontturi, E. (2017). Cellulose nanocrystals by acid vapour: towards more effortless isolation of cellulose nanocrystals. *Faraday Discussions*, 202, 315-330.

[94] Yu, H. Y., Zhang, D. Z., Lu, F. F. and Yao, J. (2016). New approach for single-step extraction of carboxylated cellulose nanocrystals for their use as adsorbents and flocculants. *ACS Sustainable Chemistry & Engineering*, 4: 2632-2643.

[95] Nicharat, A., Sapkota, J., Weder, C. and Foster, E. J. (2015). Melt processing of polyamide 12 and cellulose nanocrystals composites. *Journal of Applied Polymer Science,* 132: 42752

[96] Camarero Espinosa, S., Kuhnt, T., Foster, E. J. and Weder, C. (2013). Isolation of thermally stable cellulose nanocrystals by phosphoric acid hydrolysis. *Biomacromolecules*, 14: 1223-1230.

[97] Liu, Y., Wang, H., Yu, G., Yu, Q., Li, B. and Mu, X. (2014). A novel approach for the preparation of nanocrystalline cellulose by using phosphotungstic acid. *Carbohydrate Polymers*, 110: 415-422.

[98] Lv, D., Du, H., Che, X., Wu, M., Zhang, Y., Liu, C., Nie, S., Zhang, X. and Li, B. (2019). Tailored and integrated production of functional cellulose nanocrystals and cellulose nanofibrils via sustainable formic acid hydrolysis: kinetic study and characterization. *ACS Sustainable Chemistry & Engineering*, 7: 9449-9463.

[99] Dhar, P., Bhasney, S. M., Kumar, A. and Katiyar, V. (2016). Acid functionalized cellulose nanocrystals and its effect on mechanical, thermal, crystallization and surfaces properties of poly (lactic acid) bionanocomposites films: A comprehensive study. *Polymer*, 101: 75-92.

[100] Chen, L., Zhu, J. Y., Baez, C., Kitin, P. and Elder, T. (2016). Highly thermal-stable and functional cellulose nanocrystals and nanofibrils produced using fully recyclable organic acids. *Green Chemistry*, 18:

3835-3843.
[101] Jia, C., Bian, H., Gao, T., Jiang, F., Kierzewski, I. M., Wang, Y., ... and Hu, L. (2017). Thermally stable cellulose nanocrystals toward high-performance 2D and 3D nanostructures. *ACS Applied Materials & Interfaces*, 9: 28922-28929
[102] Bian, H., Chen, L., Dai, H. and Zhu, J. Y. (2017). Integrated production of lignin containing cellulose nanocrystals (LCNC) and nanofibrils (LCNF) using an easily recyclable di-carboxylic acid. *Carbohydrate Polymers*, 167: 167-176.
[103] Ji, H., Xiang, Z., Qi, H., Han, T., Pranovich, A. and Song, T. (2019). Strategy towards one-step preparation of carboxylic cellulose nanocrystals and nanofibrils with high yield, carboxylation and highly stable dispersibility using innocuous citric acid. *Green Chemistry*, 21: 1956-1964.
[104] Rovera, C., Ghaani, M., Santo, N., Trabattoni, S., Olsson, R. T., Romano, D. and Farris, S. (2018). Enzymatic Hydrolysis in the Green Production of Bacterial Cellulose Nanocrystals. *ACS Sustainable Chemistry & Engineering*, 6: 7725-7734.
[105] Xu, Y., Salmi, J., Kloser, E., Perrin, F., Grosse, S., Denault, J. and Lau, P. C. (2013). Feasibility of nanocrystalline cellulose production by endoglucanase treatment of natural bast fibers. *Industrial Crops and Products*, 51: 381-384.
[106] Cui, S., Zhang, S., Ge, S., Xiong, L. and Sun, Q. (2016). Green preparation and characterization of size-controlled nanocrystalline cellulose via ultrasonic-assisted enzymatic hydrolysis. *Industrial Crops and Products*, 83: 346-352.
[107] Yang, B., Dai, Z., Ding, S. Y. and Wyman, C. E. (2011). Enzymatic hydrolysis of cellulosic biomass. *Biofuels*, 2: 421-449.
[108] Satyamurthy, P., Jain, P., Balasubramanya, R. H. and Vigneshwaran, N. (2011). Prepara tion and characterization of cellulose nanowhiskers from cotton fibres by controlled microbial hydrolysis. *Carbohydrate Polymers*, 83: 122-129. 011.
[109] Zhang, Y., Lu, X. B., Gao, C., Lv, W. J., & Yao, J. M. (2012). Preparation and characterization of nano crystalline cellulose from

bamboo fibers by controlled cellulase hydrolysis. *Journal of Fiber Bioengineering and informatics*, 5: 263-271.

[110] Anderson, S. R., Esposito, D., Gillette, W., Zhu, J. Y., Baxa, U. and Mcneil, S. E. (2014). Enzymatic preparation of nanocrystalline and microcrystalline cellulose. *TAPPI JOURNAL*, 13: 35-42.

[111] Filson, P. B., Dawson-Andoh, B. E. and Schwegler-Berry, D. (2009). Enzymatic-mediated production of cellulose nanocrystals from recycled pulp. *Green Chemistry*, 11: 1808-1814.

[112] Juárez-Luna, G. N., Favela-Torres, E., Quevedo, I. R. and Batina, N. (2019). Enzymatically assisted isolation of high-quality cellulose nanoparticles from water hyacinth stems. *Carbohydrate Polymers*, 220: 110-117.

[113] Hänninen, T. and Isogai, A. (2018). Oxidative Chemistry in Preparation and Modification on Cellulose Nanoparticles. In *Nanocellulose and Sustainability*: 45-65 .CRC Press.

[114] Peyre, J., Pääkkönen, T., Reza, M. and Kontturi, E. (2015). Simultaneous preparation of cellulose nanocrystals and micron-sized porous colloidal particles of cellulose by TEMPO-mediated oxidation. *Green Chemistry*, 17: 808-811.

[115] Mikulcová, V., Bordes, R., Minařík, A. and Kašpárková, V. (2018). Pickering oil-in-water emulsions stabilized by carboxylated cellulose nanocrystals–Effect of the pH. *Food hydrocolloids*, 80: 60-67.

[116] Hirota, M., Tamura, N., Saito, T. and Isogai, A. (2010). Water dispersion of cellulose II nanocrystals prepared by TEMPO-mediated oxidation of mercerized cellulose at pH 4.8. *Cellulose*, 17: 279-288.

[117] Chen, D. and van de Ven, T. G. (2016). Morphological changes of sterically stabilized nanocrystalline cellulose after periodate oxidation. *Cellulose*, 23: 1051-1059.

[118] Bashar, M. M., Zhu, H., Yamamoto, S. and Mitsuishi, M. (2019). Highly carboxylated and crystalline cellulose nanocrystals from jute fiber by facile ammonium persulfate oxidation. *Cellulose*, 26: 3671-3684.

[119] Zhou, Y., Sun, S., Bei, W., Zahi, M. R., Yuan, Q., & Liang, H. (2018). Preparation and antimicrobial activity of oregano essential oil

Pickering emulsion stabilized by cellulose nanocrystals. *International Journal of Biological Macromolecules*, 112: 7-13.

[120] Wibowo, A., Madani, H., Judawisastra, H., Restiawaty, E., Lazarus, C. and Budhi, Y. W. (2018). An eco-friendly preparation of cellulose nano crystals from oil palm empty fruit bunches. In *IOP Conference Series: Earth and Environmental Science*, 105: 012059. IOP Publishing.

[121] Ye, S., Yu, H. Y., Wang, D., Zhu, J. and Gu, J. (2018). Green acid-free one-step hydrothermal ammonium persulfate oxidation of viscose fiber wastes to obtain carboxylated spherical cellulose nanocrystals for oil/water Pickering emulsion. *Cellulose*, 25: 5139-5155.

[122] Leung, A. C., Hrapovic, S., Lam, E., Liu, Y., Male, K. B., Mahmoud, K. A. and Luong, J. H. (2011). Characteristics and properties of carboxylated cellulose nanocrystals prepared from a novel one-step procedure. *Small*, 7: 302-305.

[123] Oun, A. A. and Rhim, J. W. (2018). Isolation of oxidized nanocellulose from rice straw using the ammonium persulfate method. *Cellulose*, 25: 2143-2149.

[124] Mascheroni, E., Rampazzo, R., Ortenzi, M. A., Piva, G., Bonetti, S. and Piergiovanni, L. (2016). Comparison of cellulose nanocrystals obtained by sulfuric acid hydrolysis and ammonium persulfate, to be used as coating on flexible food-packaging materials. *Cellulose*, 23: 779-793.

[125] Zhou, L., Li, N., Shu, J., Liu, Y., Wang, K., Cui, X., ... and Duan, Y. (2018). One-Pot Preparation of Carboxylated Cellulose Nanocrystals and Their Liquid Crystalline Behaviors. *ACS Sustainable Chemistry & Engineering*, 6: 12403-12410.

[126] Koshani, R., van de Ven, T. G., & Madadlou, A. (2018). Characterization of Carboxylated Cellulose Nanocrytals Isolated through Catalyst-Assisted H_2O_2 Oxidation in a One-Step Procedure. *Journal of Agricultural and Food Chemistry*, 66(29): 7692-7700.

[127] Novo, L. P., Bras, J., García, A., Belgacem, N., & Curvelo, A. A. (2015). Subcritical water: a method for green production of cellulose

nanocrystals. *ACS Sustainable Chemistry & Engineering*, 3: 2839-2846.

[128] Park, N. M., Choi, S., Oh, J. E. and Hwang, D. Y. (2019). Facile extraction of cellulose nanocrystals. *Carbohydrate Polymers*, 223: 115114.

[129] Sen, S., Martin, J. D. and Argyropoulos, D. S. (2013). Review of cellulose non-derivatizing solvent interactions with emphasis on activity in inorganic molten salt hydrates. *ACS Sustainable Chemistry & Engineering*, 1: 858-870.

[130] Tan, X., Li, X., Chen, L. and Xie, F. (2016). Solubility of starch and microcrystalline cellulose in 1-ethyl-3-methylimidazolium acetate ionic liquid and solution rheological properties. *Physical Chemistry Chemical Physics*, 18: 27584-27593.

[131] Acharya, S., Hu, Y. and Abidi, N. (2018). Mild condition dissolution of high molecular weight cotton cellulose in 1-butyl-3-methylimidazolium acetate/N, N-dimethylacetamide solvent system. *Journal of Applied Polymer Science*, 135: 45928.

[132] Tan, X. Y., Hamid, S. B. A. and Lai, C. W. (2015). Preparation of high crystallinity cellulose nanocrystals (CNCs) by ionic liquid solvolysis. *Biomass and Bioenergy*, 81: 584-591.

[133] Abushammala, H., Krossing, I. and Laborie, M. P. (2015). Ionic liquid-mediated technology to produce cellulose nanocrystals directly from wood. *Carbohydrate Polymers*, 134: 609-616.

[134] Cantero, D. A., Bermejo, M. D. and Cocero, M. J. (2015). Governing chemistry of cellulose hydrolysis in supercritical water. *ChemSusChem*, 8: 1026-1033.

[135] Novo, L. P., Bras, J., García, A., Belgacem, N. and da Silva Curvelo, A. A. (2016). A study of the production of cellulose nanocrystals through subcritical water hydrolysis. *Industrial crops and products*, 93: 88-95

[136] Domingues, R. M., Gomes, M. E. and Reis, R. L. (2014). The potential of cellulose nanocrystals in tissue engineering strategies. *Biomacromolecules*, 15: 2327-2346.

[137] Sun, H., Xu, Z. and Gao, C. (2013). Multifunctional, ultra-flyweight,

synergistically assembled carbon aerogels. *Advanced Materials*, 25: 2554-2560.

[138] Abitbol, T., Rivkin, A., Cao, Y., Nevo, Y., Abraham, E., Ben-Shalom, T., Lapidot, S. and Shoseyov, O. (2016). Nanocellulose, a tiny fiber with huge applications. *Current Opinion in Biotechnology*, 39: 76-88.

[139] Endes, C., Mueller, S., Kinnear, C., Vanhecke, D., Foster, E. J., Petri-Fink, A., Weder, C., Clift, M. J. D. and Rothen-Rutishauser, B. (2015). Fate of cellulose nanocrystal aerosols deposited on the lung cell surface in vitro. *Biomacromolecules*, 16: 1267-1275.

[140] Mahmoud, K. A., Mena, J. A., Male, K. B., Hrapovic, S., Kamen, A. and Luong, J. H. (2010). Effect of surface charge on the cellular uptake and cytotoxicity of fluorescent labeled cellulose nanocrystals. *ACS Applied Materials & Interfaces*, 2: 2924-2932.

[141] Wang, Y. and Chen, L. (2014). Cellulose nanowhiskers and fiber alignment greatly improve mechanical properties of electrospun prolamin protein fibers. *ACS Applied Materials & Interfaces*, 6: 1709-1718.

[142] He, X., Xiao, Q., Lu, C., Wang, Y., Zhang, X., Zhao, J., Zhang, W., Zhang, X. and Deng, Y. (2014). Uniaxially aligned electrospun all-cellulose nanocomposite nanofibers reinforced with cellulose nanocrystals: scaffold for tissue engineering. *Biomacromolecules*, 15: 618-627.

[143] Rezwan, K., Chen, Q. Z., Blaker, J. J. and Boccaccini, A. R. (2006). Biodegradable and bioactive porous polymer/inorganic composite scaffolds for bone tissue engineering. *Biomaterials*, 27: 3413-3431.

[144] Shin, H., Jo, S., & Mikos, A. G. (2003). Biomimetic materials for tissue engineering. *Biomaterials*, 24(24), 4353-4364.

[145] Williams, D. (2004). Benefit and risk in tissue engineering. *Materials Today*, 7: 24-29.

[146] Griffith, L. G. (2002). Emerging design principles in biomaterials and scaffolds for tissue engineering. *Annals of the New York Academy of Sciences*, 961: 83-95.

[147] Lee, S. J., Lim, G. J., Lee, J. W., Atala, A. and Yoo, J. J. (2006). In vitro evaluation of a poly (lactide-co-glycolide)–collagen composite

scaffold for bone regeneration. *Biomaterials*, 27: 3466-3472.

[148] Mikos, A. G. and Temenoff, J. S. (2000). Formation of highly porous biodegradable scaffolds for tissue engineering. *Electronic Journal of Biotechnology*, 3: 23-24.

[149] Jorfi, M. and Foster, E. J. (2015). Recent advances in nanocellulose for biomedical applications. *Journal of Applied Polymer Science*, 132: 41719.

[150] Grishkewich, N., Mohammed, N., Tang, J. and Tam, K. C. (2017). Recent advances in the application of cellulose nanocrystals. *Current Opinion in Colloid & Interface Science*, 29: 32-45.

[151] Kumar, A., Negi, Y. S., Choudhary, V. and Bhardwaj, N. K. (2014). Microstructural and mechanical properties of porous biocomposite scaffolds based on polyvinyl alcohol, nano-hydroxyapatite and cellulose nanocrystals. *Cellulose*, 21: 3409-3426.

[152] Shaheen, T. I., Montaser, A. S. and Li, S. (2019). Effect of cellulose nanocrystals on scaffolds comprising chitosan, alginate and hydroxyapatite for bone tissue engineering. *International Journal of Biological Macromolecules*, 121: 814-821.

[153] Li, W., Guo, R., Lan, Y., Zhang, Y., Xue, W. and Zhang, Y. (2014). Preparation and properties of cellulose nanocrystals reinforced collagen composite films. *Journal of Biomedical Materials Research Part A*, 102: 1131-1139.

[154] Abraham, E., Kam, D., Nevo, Y., Slattegard, R., Rivkin, A., Lapidot, S. and Shoseyov, O. (2016). Highly modified cellulose nanocrystals and formation of epoxy-nanocrystalline cellulose (CNC) nanocomposites. *ACS Applied Materials & Interfaces*, 8: 28086-28095.

[155] Abraham, E., Weber, D. E., Sharon, S., Lapidot, S. and Shoseyov, O. (2017). Multifunctional cellulosic scaffolds from modified cellulose nanocrystals. *ACS Applied Materials & Interfaces*, 9: 2010-2015.

[156] Schyrr, B., Pasche, S., Voirin, G., Weder, C., Simon, Y. C. and Foster, E. J. (2014). Biosensors based on porous cellulose nanocrystal–poly (vinyl alcohol) scaffolds. *ACS Applied Materials & Interfaces*, 6:12674-12683.

[157] Dong, S. and Roman, M. (2007). Fluorescently labeled cellulose nanocrystals for bioimaging applications. *Journal of the American Chemical Society*, 129: 13810-13811.

[158] Pilla, S. (Ed.). (2011). *Handbook of Bioplastics and Biocomposites Engineering Applications*. John Wiley & Sons.

[159] Mariano, M., El Kissi, N. and Dufresne, A. (2014). Cellulose nanocrystals and related nanocomposites: review of some properties and challenges. *Journal of Polymer Science Part B: Polymer Physics*, 52: 791-806.

[160] Kalia, S., Dufresne, A., Cherian, B. M., Kaith, B. S., Avérous, L., Njuguna, J. and Nassiopoulos, E. (2011). Cellulose-based bio-and nanocomposites: a review. *International Journal of Polymer Science*, 2011: 837875.

[161] Favier, V., Chanzy, H. and Cavaille, J. Y. (1995). Polymer nanocomposites reinforced by cellulose whiskers. *Macromolecules*, 28: 6365-6367.

[162] Kowalczyk, M., Piorkowska, E., Kulpinski, P. and Pracella, M. (2011). Mechanical and thermal properties of PLA composites with cellulose nanofibers and standard size fibers. *Composites Part A: Applied Science and Manufacturing*, 42: 1509-1514.

[163] Marcovich, N. E., Auad, M. L., Bellesi, N. E., Nutt, S. R. and Aranguren, M. I. (2006). Cellulose micro/nanocrystals reinforced polyurethane. *Journal of Materials Research*, 21: 870-881.

[164] Angles, M. N. and Dufresne, A. (2001). Plasticized starch/tunicin whiskers nanocomposite materials. 2. Mechanical behavior. *Macromolecules*, 34: 2921-2931.

[165] Chauve, G., Heux, L., Arouini, R. and Mazeau, K. (2005). Cellulose poly (ethylene-co-vinyl acetate) nanocomposites studied by molecular modeling and mechanical spectroscopy. *Biomacromolecules*, 6: 2025-2031.

[166] Chazeau, L., Cavaille, J. Y., Canova, G., Dendievel, R. and Boutherin, B. (1999). Viscoelastic properties of plasticized PVC reinforced with cellulose whiskers. *Journal of Applied Polymer Science*, 71: 1797-1808.

[167] Fortunati, E., Luzi, F., Janke, A., Häußler, L., Pionteck, J., Kenny, J. M. and Torre, L. (2017). Reinforcement effect of cellulose nanocrystals in thermoplastic polyurethane matrices characterized by different soft/hard segment ratio. *Polymer Engineering & Science*, 57: 521-530.

[168] Lu, Y., Weng, L. and Cao, X. (2006). Morphological, thermal and mechanical properties of ramie crystallites—reinforced plasticized starch biocomposites. *Carbohydrate Polymers*, 63: 198-204.

[169] Zoppe, J. O., Peresin, M. S., Habibi, Y., Venditti, R. A. and Rojas, O. J. (2009). Reinforcing poly (ε-caprolactone) nanofibers with cellulose nanocrystals. *ACS Applied Materials & Interfaces*, 1: 1996-2004.

[170] Paralikar, S. A., Simonsen, J. and Lombardi, J. (2008). Poly (vinyl alcohol)/cellulose nanocrystal barrier membranes. *Journal of Membrane Science*, 320: 248-258.

[171] Mohammadkazemi, F., Doosthoseini, K., Ganjian, E. and Azin, M. (2015). Manufacturing of bacterial nano-cellulose reinforced fiber–cement composites. *Construction and Building Materials*, 101: 958-964.

[172] Cao, Y., Zavaterri, P., Youngblood, J., Moon, R. and Weiss, J. (2015). The influence of cellulose nanocrystal additions on the performance of cement paste. *Cement and Concrete Composites*, 56: 73-83.

[173] Lange, J. and Wyser, Y. (2003). Recent innovations in barrier technologies for plastic packaging—a review. *Packaging Technology and Science: An International Journal*, 16: 149-158.

[174] Yang, H., Tejado, A., Alam, N., Antal, M. and van de Ven, T. G. (2012). Films prepared from electrosterically stabilized nanocrystalline cellulose. *Langmuir*, 28: 7834-7842.

[175] George, J. and Sabapathi, S. N. (2015). Cellulose nanocrystals: synthesis, functional properties, and applications. *Nanotechnology, Science and Applications*, *8*, 45-54.

[176] Ma, H., Burger, C., Hsiao, B. S. and Chu, B. (2011). Ultrafine polysaccharide nanofibrous membranes for water purification. *Biomacromolecules*, 12: 970-976.

[177] Ching, Y. C., Ali, M. E., Abdullah, L. C., Choo, K. W., Kuan, Y. C.,

Julaihi, S. J., Chuah, C. H. and Liou, N. S. (2016). Rheological properties of cellulose nanocrystal-embedded polymer composites: a review. *Cellulose*, 23: 1011-1030.

[178] Jiang, E., Amiralian, N., Maghe, M., Laycock, B., McFarland, E., Fox, B., Martin, D. J. and Annamalai, P. K. (2017). Cellulose nanofibers as rheology modifiers and enhancers of carbonization efficiency in polyacrylonitrile. *ACS Sustainable Chemistry & Engineering*, 5: 3296-3304.

[179] Yang, J., Han, C. R., Duan, J. F., Xu, F. and Sun, R. C. (2013). Mechanical and viscoelastic properties of cellulose nanocrystals reinforced poly (ethylene glycol) nanocomposite hydrogels. *ACS Applied Materials & Interfaces*, 5: 3199-3207.

[180] Oguzlu, H., Danumah, C. and Boluk, Y. (2016). The role of dilute and semi-dilute cellulose nanocrystal (CNC) suspensions on the rheology of carboxymethyl cellulose (CMC) solutions. *The Canadian Journal of Chemical Engineering*, 94: 1841-1847.

[181] Li, M. C., Wu, Q., Song, K., Qing, Y. and Wu, Y. (2015). Cellulose nanoparticles as modifiers for rheology and fluid loss in bentonite water-based fluids. *ACS Applied Materials & Interfaces*, 7: 5006-5016.

[182] Hill, R. J. (2008). Elastic modulus of microfibrillar cellulose gels. *Biomacromolecules*, 9: 2963-2966.

[183] Yang, X., Shi, K., Zhitomirsky, I. and Cranston, E. D. (2015). Cellulose nanocrystal aerogels as universal 3D lightweight substrates for supercapacitor materials. *Advanced Materials*, 27: 6104-6109.

[184] Chen, G., Chen, T., Hou, K., Ma, W., Tebyetekerwa, M., Cheng, Y., Weng, W. and Zhu, M. (2018). Robust, hydrophilic graphene/ cellulose nanocrystal fiber-based electrode with high capacitive performance and conductivity. *Carbon*, 127: 218-227.

[185] Bokhari, S. W., Hao, Y., Siddique, A. H., Ma, Y., Imtiaz, M., Butt, R., Hui, P., Li, Y. and Zhu, S. (2019). Assembly of hybrid electrode rGO–CNC–MnO_2 for a high performance supercapacitor. *Results in Materials*, 1: 100007.

[186] Yuan, H., Pan, H., Meng, X., Zhu, C., Liu, S., Chen, Z., Ma, J. and

Zhu, S. (2019). Assembly of MnO/CNC/rGO fibers from colloidal liquid crystal for flexible supercapacitors via a continuous one-process method. *Nanotechnology*, 30:465702.

[187] Zhou, Y., Fuentes-Hernandez, C., Khan, T. M., Liu, J. C., Hsu, J., Shim, J. W., Dindar, A., Youngblood, J. P., Moon, R. J. and Kippelen, B. (2013). Recyclable organic solar cells on cellulose nanocrystal substrates. *Scientific reports*, 3:1536.

[188] Zhou, Y., Khan, T. M., Liu, J. C., Fuentes-Hernandez, C., Shim, J. W., Najafabadi, E., Youngblood, J. P., Moon, R. J. and Kippelen, B. (2014). Efficient recyclable organic solar cells on cellulose nanocrystal substrates with a conducting polymer top electrode deposited by film-transfer lamination. *Organic Electronics*, 15: 661-666.

[189] Yuwawech, K., Wootthikanokkhan, J., Wanwong, S. and Tanpichai, S. (2017). Polyurethane/esterified cellulose nanocrystal composites as a transparent moisture barrier coating for encapsulation of dye sensitized solar cells. *Journal of Applied Polymer Science*, 134: 45010.

[190] Sadasivuni, K. K., Kafy, A., Zhai, L., Ko, H. U., Mun, S., and Kim, J. (2015). Transparent and flexible cellulose nanocrystal/reduced graphene oxide film for proximity sensing. *Small*, 11: 994-1002.

[191] Najafabadi, E., Zhou, Y. H., Knauer, K. A., Fuentes-Hernandez, C. and Kippelen, B. (2014). Efficient organic light-emitting diodes fabricated on cellulose nanocrystal substrates. *Applied Physics Letters*, 105: 063305.

In: Cellulose Nanocrystals
Editor: Orlene Croteau

ISBN: 978-1-53616-747-4

Chapter 2

REVIEW ON: CURRENT RESEARCH APPLICATIONS OF CELLULOSE NANOCRYSTALS, ITS SOURCES, FABRICATION METHODS, AND PROPERTIES

***Le Van Hai*[1,3,*], *Srikanth Narayanan*[2], *Ruth M. Muthoka*[1] *and Jaehwan Kim*[1,†]**

[1]CRC for Nanocellulose Future Composites, Inha University, Incheon City, Republic of Korea

[2]Applied Research and Innovation, New Brunswick Community College, Saint John, Canada

[3]Department for Management of Science and Technology Development, Ton Duc Thang University, Ho Chi Minh, Vietnam

[*]Corresponding Author's Email: levanhai121978@gmail.com; levanhai@tdtu.edu.vn.
[†]Corresponding Author's Email: jaehwan@inha.ac.kr.

ABSTRACT

In the 21st century, cellulose based nano-sized materials will be regarded as an innovative matter that encouraging production of high value-added pulp and paper products, and provide solutions for technological innovations for various applications. Recently, there is a wide range of cellulose based nano-materials that have been extracted, and developed for various applications. The current research interest is towards conducting a large scale research and development activities on numerous applications of nano-materials for commerical use. Cellulose nanofiber is a eco-friendly materials with a potential applications as reinforced filler in various composites, targeted drug delivery, smart materials, 3D printing, and automotive interior parts. The first cellulose nanofiber or micelles was chemically (acid hydrolysis) extracted by Ranby in 1949. Based on fabrication techniques, cellulose nanofibers can be categorized into cellulose nanocrystals, nanofibrillated celulose and TEMPO cellulose nanofibers. Cellulose nanofibers can be extracted by top down and bottom up approaches such as mechanical, enzymatic hydrolysis, TEMPO, acid hydrolysis and production of bacterial cellulose respectively. In recent years, the use of cellulose nanocrystals for different applications have gained much research interest because they are bio-degradable, bio-compatible, carbon neutral, and readily available from renewable resources. In the acid hydrolysis process, the cellulose nanocrystals (CNCs) amorphous regions are destroyed while the crystalline domains are kept intact. In this review chapter, the recent developments on research and applications of cellulose nanocrystal extracted by acid hydrolysis will be discussed. This chapter aims to provide a detailed description on i) Cellulose nanocrystals isolation methods, characterization, and properties ii) Current and future trends in terms of development of CNCs for various applications and iii) Concluding remarks on cellulose nanocrystals major use in advanced engineering applications such as composites, smart materials, electronic devices, automobile interiors, surface coating, used as a lubricant, transparent films, targeted drug delivery and etc.

Keywords: cellulose nanocrystal, nanocellulose, green materials, composite materials, bio-degradable materials

INTRODUCTION

Wood cellulose as a green material has the characteristics such as biodegradable, renewable, recyclable and reusable. Traditionally, cellulose is used for diverse applications such as furniture, construction, papers, food packaging, and cosmetics. According to Ummartyotin and Manuspiya 2015 [1], Hai et al. 2019 [2] cellulose was discovered by Anselme Payen, a French chemist in 1839. Such cellulose microfibers are extensively used in production of writing and printing papers, paper-based packaging for food and other applications. Cellulose nanofiber is a new class of nano-sized materials which has a dimensions of 10^{-9} m. With aunique characterisitcs of pristine crystalline structure, 3D network formation, and thixotrophy properties that are vastly appreciated by academic researcher and the industry partners as well. The first cellulose nanomaterials, called as micelle and microcrystalline cellulose, was discovered by Ranby in 1949 [2]. After that the next cellulose nanofibrillated fiber was prepared by mechnical process and reported by Turbark in 1983 [2]. Since then, the methods of isolating cellulose nanofibers have been extensively explored by different research groups. In recent years, a large patent's database search returned results on CNCs that showed cellulose nanofibers research have gained much attention becasue of its potential applications in the automotive sector, chemical sensors, flexible screen, targeted drug release, and 3D printing.

In general, nanocellulose isolation approaches can be classified as:

Nanofibrillated Cellulose (NFCs) are mainly isolated by mechanical processes such as grinder, homogenizer, microfluidizer, ballmill and mass colliders.

TEMPO-oxidation process produces cellulose nanofibrils namely TEMPO-oxidized cellulose nanofiber. TEMPO-oxidation cellulose nanofiber was introduced by Isogai group of Japan, the main contributor for research and applications of TEMPO-oxidized cellulose nanofiberas a reinforced filler in the autmotive interior parts and in the food industry as a additives.

Another form of bio-degrdabale cellulose nanofibers is termed as Bacterial Celluloses (BCs) which are synthesized by bacteria.

Also, cellulose nanocrystals can be prepared by enzymatic treatment combined with mechanical methods is practiced.

In the acid hydrolysis, the isolation of nanocrystals cellulose (CNCs) from cellulose pulp is accomplished by using mineral acids such as HCl, H_2SO_4, HNO_3...

Cellulose nanofiber prepared from different sources and isolation methods resulted in different lengths, width, crystallinity, and thermal degradation properties. Owing to numerous publications on different types of nanocellulose isolation methods, this chapter will mainly focus on the cellulose nanocrystals that are produced by acid hydrolysis only.

As described by different research groups, cellulose nanocrystal, cellulose nano-whisker, or also nanocrystalline cellulose are same nomenclature provided form a nanocelluloses fabricated using acid hydrolysis methods. In this chapter, the brief description on nanocrystals, nanocrystalline cellulose, or nano-whisker cellulose fabrication techniques, yield and composite properties will be discussed. Based on our literature reviews, cellulose nanocrystal havea width of ~ 3 to 30 nm which depends on the cellulose (rawmaterial) sources and isolation conditions. Based on fabrication methods, the length of cellulose nanocrystals could vary from a hundred nanometers to microns. In the special case, the isolated cellulose nanocrystals from marine invertebrate animal such as tunicate have a length more than a micron as reported by Moon et al. 2011 [3]. Also, cellulose nanocrystals prepared via acid hydrolysis showed a superior mechanical properties interms of tensile and elastic modulus compared to other carbon based nano materials such as Kevlar-49 fiber, and carbon fibers. According to Moon et al. 2011 [3], cellulose nanocrystal has a tensile strength of about 7.5 to 7.7 GPa and elastic modulus in the axial direction could be 110 to 220 GPa. Nanocellulose mechanical properties are better compared to Kevlar tensile strength and elastic modulus of 3.5 ± 2 and 127 ± 3 GPa respectively.

Furthermore, CNCs unique characteristics of self-assembly, iridescent colors and distinct nematic pitch has potential applications in different areas of chemical sensors, various medical devices such as blood diagnostic strips and targeted drug release, and bank note security features as reported by Dumanli et al. 2016 [4]. CNC properties of self-assembly and iridescent color are exhibited under certain conditions, due to the chiral nematic ordering and network formation as reported by Habibi et al. 2010 [5], Picard et al. 2012 [6], Gray and Mu 2015 [7]. In addition, the self-ordering and 3D network formation properties of nanomaterials are mainly affected by various factors such as hydrogen bond, balance force, hydrophobic and hydrophilic forces, colloidal interfacial interaction, and heating or cooling [8, 9, 10].

Furthermore, the study on CNCs used as reinforced fillers for different bio-nanocomposites are very exhaustive. The literature review mostly discusses the reinforcement of CNCs into polymer matrix to improve the composites mechanical properties. Figure 1 shows a brief overview of the conversion process of cellulose into cellulose nanocrystals by acid hydrolysis.

In general, the cellulose nanofibers can be categorized into celulose nanocrystal, nanofibrillated, and TEMPO-oxidized cellulose nanofiber. The nanofibrillated cellulose is flexible as it retains both amophorous and crystalline domains, and vice versa the nanocrystal cellulose is very stiff with crystalline regions remaining intact after acid hydrolysis, thus CNCs are highly crystalline. TEMPO-oxidized cellulose nanofiber is very flexible but cellulose nanocrystals are rod liked structure, short and brittle. The nano cellulose crystallinity mainly depends on the use of cellulose raw material sources, treatment conditions and chemical dosage. The cellulose nanocrystals extracted from recycled fiber or use of high concentration acid treatment incurs degradation of the amorphous regions along with crystalline structure thus low crystallinity CNCs are obtained. The illustration of morphology of nanofibrillated cellulose, TEMPO-oxidized cellulose nanofiber and nanocrystal celluloses are exhibited in Figure 2.

CELLULOSE NANOCRYSTAL SOURCES AND FABRICATION METHODS

Cellulose Nanocrystals Sources

Cellulose, a basic constituent of plant cells was first discovered by Anselme Payen in 1839 [1-2]. Cellulose is a linear chain polysaccharide with a repeating glucose units linked by β-1-4 glycosidic bonds. Cellulose is used in the different form of paper products that is well-known such as tissue papers, notebooks, bank notes, and wallpapers. Cellulose is mainly extracted from wood sources, but it also can be produced from non-wood sources such as bagasse, straw, rice stem, red algae, cattail, bacterial cellulose, and many others. Table 1 illustrates example of cellulose nanocrystals isolated from wood and non-wood cellulose sources and their potential applications of CNCs.

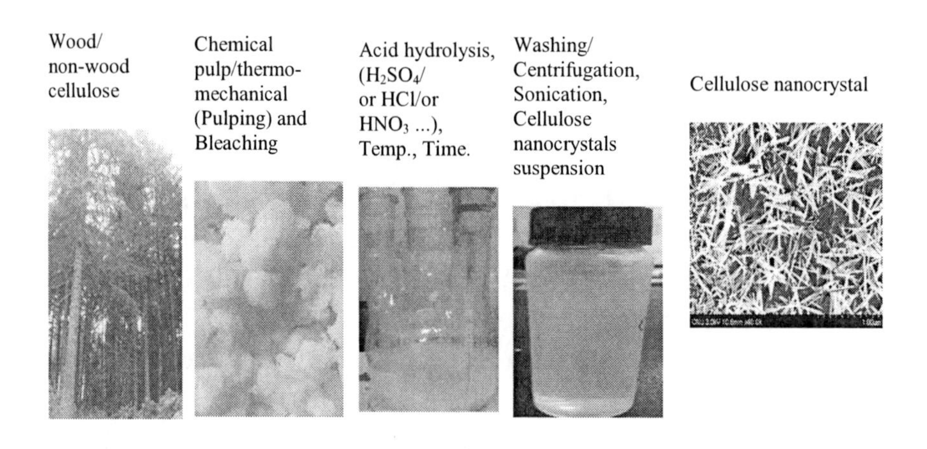

Figure 1. A brief process overview of conversion cellulose to nanocrystal fiber.

In order to isolate cellulose nanocrystal via chemical route, the types of acid used are H_2SO_4, HCl, HNO_3 and others. The types of cellulose source, chemical concentration, reaction time and temperature used in the isolation of cellulose nanocrystals are illustrated in Table 2. The first extraction of cellulose nanocrystals using acid hydrolysis was reported by Ranby in 1949 [11]. Such isolated cellulose nanocrystals were called micelle and sometimes as cellulose microcrystals. Cellulose nanocrystals had dimensions of width less than 100 nm and length of hundred nm to micrometers respectively. The cellulose nanocrystal dimensions interms of shape and size mainly depends on fabrication methods, cellulose sources, and reaction conditions.

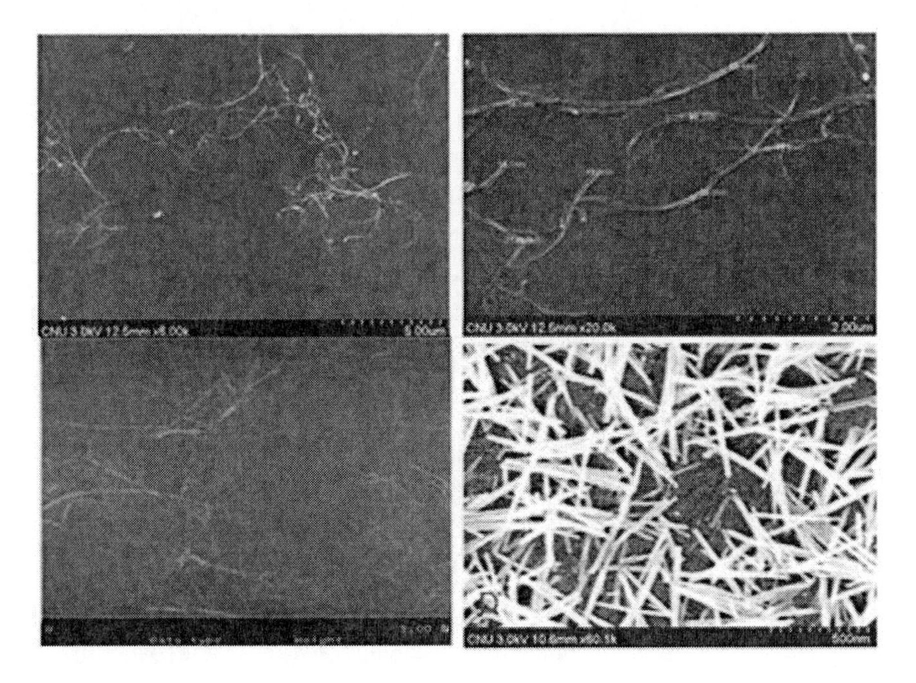

Figure 2. Nanofibrillated cellulose (A and B/FE-SEM) and TEMPO-oxidized cellulose nanofiber (C/AFM) and cellulose nanocrystals (D/FE-SEM).

In the acid hydrolysis extraction, the concentration of H_2SO_4 are varied from 55 to 65%, reaction temperature as low as room temperature up to 100°C, and reaction time (~15 minutes to hours) which depends on the chemical concentration and reaction temperature.

From the reporting of CNCs isolated using H_2SO_4, the researcher altered the reaction conditions such as use of different acid types, acid concentrations, reaction times and temperatures to produce a consistent CNCs quality with higher yields. Due to the necessity of strict shape and size requirements for specific use of CNCs in composite applications, the researcher mainly focused on optimizing reaction conditions to produce consistent cellulose nanocrystals thereby controlling the reaction parameters or cellulose sources. Luzi et al. 2014 [49] reported that the optimized conditions to produce CNC from Carmagnola hemp fiber was accomplished by prior chemical pre-treatment methods followed by acid hydrolysis.

Table 1. Wood and non-wood fiber to produce nanocrystals cellulose and applications

Wood/Non-wood cellulose sources	Applications	References
Bamboo	Starch composites	[12]
Wood, cotton, Cattail, red algae	All-cellulose nanocomposites	[13]
Cotton	Gate dielectric, transistors	[14]
Sugarcane bagasse	Functional hybrid nanofiller	[15]
Ramie fiber	All-cellulose nanocomposites	[16]
MCC	PHB packaging	[17]
Wastepaper	PET packaging	[18]
Soy hulls	Extraction nanocrystals cellulose	[19]
Rice straw	Extraction nanocrystals cellulose	[20]
Tunicate	Nanocomposite polymer electrolytes	[21]
Grass	Reinforcement in poly (lactic acid)	[22]
MCC	Reinforcement in commercial silicone rubber	[23]

Another work from Lu et al. 2016 [50], demonstrated CNC extraction yields with better product quality were achieved by controlling the experiment parameters such as acid concentrations, reaction times, and ball milling power intensity were reported. Also, high yield CNC of 88% were extracted by the method of mechanochemical activation followed by phosphotungstic acid hydrolysis. In a recent paper by Garcıa et al. 2018 [51], the isolation of CNCs from pinecone using the reaction conditions of sulfuric acid concentrations of 65 wt.%, T = 45°C, and holding periods for 30, 45 and 90 minutes.

Table 2. Fabrication conditions, cellulose sources and L/W ratio

Cellulose sources	Concentration, temperature and Time	L (nm)	W (nm)	Aspect ratio	Thermal decomposition (Cellulose/NCC) °C	References
Sulfuric acid (H_2SO_4)						
Wood, cotton	2.5N, Boiling, 60 to 480 mins				NB	[24]
China Cotton, Africa Cotton, waste tissue,	47%, 60°C, 120 mins	50-200	10-90	~2-20	338, 290 and 353/360, 358, and 367	[25]
Coconut husk	64%, 45°C, 120, 150 and 180 mins	177-218	5.3-6.6	~35-44	200/120	[26]
Wheat Straw cellulose	65%, 25°C, 60 mins	150-300	5	~30-60	NB	[27]
Softwood pulp	64%, 45°C, 45 mins	NB	NB	NB	NB	[28]
Bleached aspen kraft pulp, Old newsprint deinked pulp	64%, 45 to 55°C, and 25 to 30 mins	NB	NB	NB	NB	[29]
Oil palm empty fruit bunches	64%, 45°C, 15 to 90 mins	100-2000	~2	~50-1000	250/190, 225, 230, 240	[30]
Sugarcane bagasse	64%, 45°C, 60 mins	250-480	20-60	~ 10	270-363/ 236-300	[31]
Wastepaper	60% (v/v), 45°C, 60 mins	100-300	3-10	~10	NB	[32]
Kenaf blast fiber	65%, 45°C, 20-120 mins	124-166	12-13	~11-13	353/198-358	[33]
Capim mombaca	11.22M, 40°C, 10-40 mins,	30-320	3-6	~25-50	281-331/ 235-312	[34]
Micro-crystalline cellulose	64 wt.%, 50°C, 60 mins	230	20	~11	NB	[35]
Hardwood	64 wt.%, 50°C, 50 mins	202	15	~13	NB	[10]
Hardwood Softwood, Cotton, Cattail, and Red algae	64 wt.%, 50°C, 50 mins	171,179, 278, 249 and 432	15, 17, 33, 18, 29	~8-15	290-360/ 195-250	[13]
Pineapple leaf	9.17M, 45°C, varies from 5 to 60 mins	249	4.5	~55	NB	[36]

Table 2. (Continued)

Cellulose sources	Concentration, temperature and Time	L (nm)	W (nm)	Aspect ratio	Thermal decomposition (Cellulose/NCC) °C	References
Sulfuric acid (H_2SO_4)						
Pineapple Peel	65 wt.% of H2SO4, 55°C, from 20, 35, 50, 60 and 75 mins	NB	18-25	NB	NB	[37]
Oil palm fruit bund	64% H_2SO_4, 40°C for 60 min	>100	10-20	~5-10	275-326/ 125-418	[38]
Bamboo	46%, 55°C, 10-50 mins	200-500	<20	~10-50	NB	[39]
North African Grass	64% w/w, 45°C, 30 mins	180-238	21-32	~9-10	270-430/ 285-353	[40]
Rice husk	10.0 M, 50°C, 40 mins	NB	15-20	~10-15	NB	[41]
Hydrochloric acid (HCl)						
Cotton or wood cellulose	2,5N Boiling, 1-8 hours	NB	NB	NB		[11]
Cotton (MCC)	2.5 N, 85°C, 180mins	188	19	~10	280-380/ 170-310	[14]
Whatman Paper #4	3N, 90°C, 120 mins	256	242	~1	NB	[42]
MCC	4M, 110°C, 180mins	168-245	10-23	~11-17	NB	[43]
MCC	6M, 110°C, 180 mins	230	16	~20	339-359/ 235-364	[44]
Cotton gin motes	4N, 70°C, 150mins				268-387/ 202-422	[45]
Other acid hydrolysis methods						
Whatman #1 filter paper	1.5, 2.5, or 4.0 M HBr, 100°C, 1, 2, 3, or 4 hrs	100-400	7.0-8.6	~15-45	NB	[46]
Cotton microcrystalline cellulose (MCC)	4-8.0 M $H_3PW_{12}O_{40}$, 110°C, 3hrs	200-300	8-10	~20-30	340-390/ 275 - 430	[47]
Hardwood	50-85%, $H_3PW_{12}O_{40}$, 110°C, 15-30hrs	>100	15-40	~3-7	300-370/ 170-350	[48]

NB: not available.

Properties of Cellulose Nanocrystals

Morphology

The morphology of nanocrystals can be investigated by different techniques such as TEM, FE-SEM, AFM, and DSL respectively. For the determination of the fiber morphology, the sample prepared for SEM, FE-SEM requires a surface coating with different substrates such as Pt or Pb. For TEM analysis, the samples were laid on the carbon gird and allowed to air dry before measurement can be done; but also, CNCs can be coated with different substrates to obtain a clear morphology images of CNCs. In the case of AFM study, the samples were sprayed on various substrates such as silicon wafer, glass substrate or others to get a uniform distribution of single fiber analysis.

With respect to the morphology of the nanocrystals, an extensive collection of data in the literature are available. Depending on the raw material sources and fabrication methods, the dimensions such as width and length of nanocrystals could vary in between 3 to 30 nm and 100 nm to microns respectively. Typically, the length of the nanocrystals cellulose isolated (acid hydrolysis) from wood as a source are around 100 to 500 nm [27, 3]. Table 2 illustrates the reaction conditions and aspect ratio (length/width) of cellulose nanocrystals prepared from various cellulose sources and fabrication methods. Figure 3 shows the morphology of CNCs viewed under AFM and SEM. For example, the dimensions of CNCs can be measured by different image anlaysis software, as shown in Figure 3, the length and width dimensions of CNCs are analyzed from SEM image using ImageJ.

Figure 3 shows that the length of CNCs is around 200 to 500 nm and exhibiting a rod-liked structure. SEM sample was prepared by laying CNCs over a mica disk and coated with Pt. For more precision measurement, CNC width were recorded at a magnification of 500 nm to 1 μm. In other case, CNC dimensions were extracted from AFM image via profile measurement.

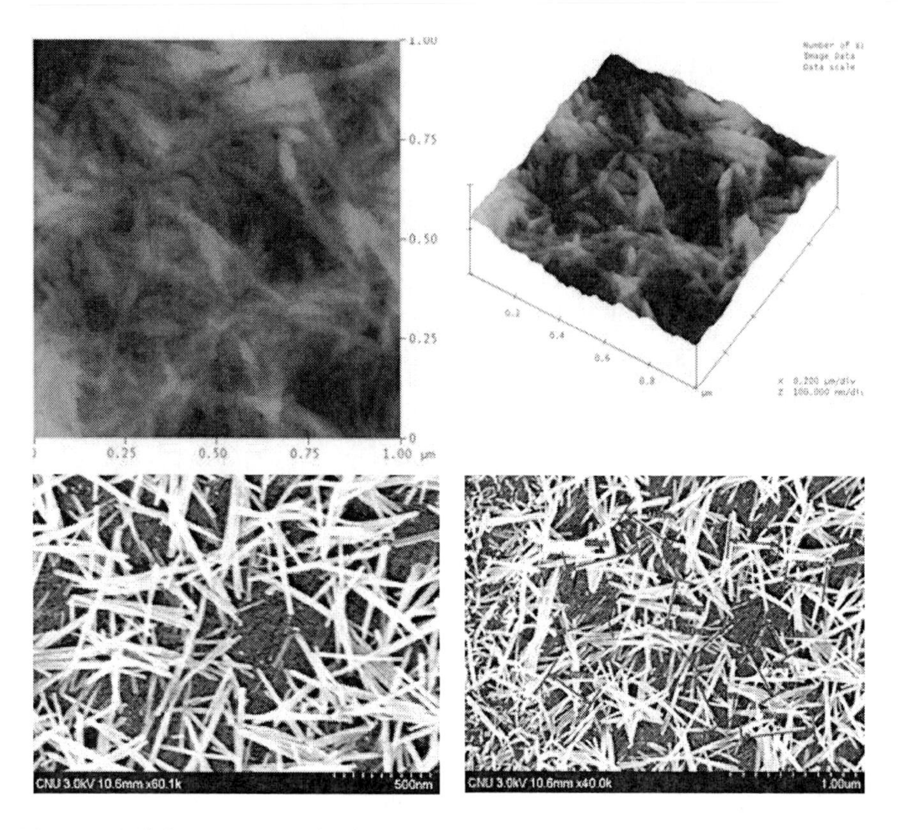

Figure 3. Cellulose nanocrystals viewed under AFM and FE-SEM.

Thermal Properties

In general, the thermal degradation of cellulose nanocrystals is lower than cellulose microfibers due to the presence anion charges on the surface. With surface modifications, the cellulose nanocrystals can retain their original thermal stability or even higher thermal properties than cellulose microfibers is possible. One of the most important properties of reinforced filler for composite applications is the requirement of higher thermal stability of CNCs. At the first stage of thermal decomposition, the water evaporation appeared in the range around 100°C to 120°C. The weight loss accounted of evaporation is around 6% to 10% for both native cellulose and cellulose nanocrystal as shown in Figure 4. CNCs thermal decomposition occurred at around 300°C and reached a pleateau at 360°C. As the figure shows that the onset temperature for cellulose is around 360°C. In the case of cellulose nanocrystal, the TG/DTG shows two thermal degradation peaks were observed. At first, cellulose nanocrystal degraded in between 200°C to 250°C and then followed by another thermal degradation occurred at 350-360°C. In order to retain the thermal stability of CNCs, several research groups have proposed CNCs neutralization with caustic soda. In general, the thermal decomposition of cellulose is much higher compared to cellulose nanocrystals as shown in Table 2.

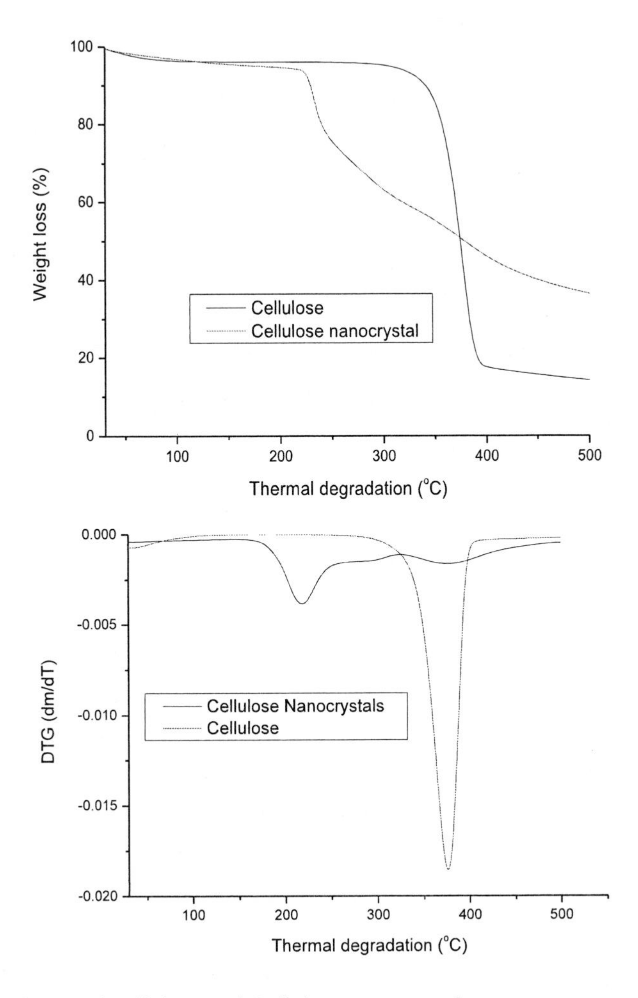

Figure 4. TG/DTG of Cellulose and Cellulose nanocrystals.

Crystallinity Index

Crystallinity is another important parameter that plays a key role in evaluating strength of reinforced filler materials. Higher crystallinity indicates enhanced mechanical properties of materials such as tensile strength and elastic modulus. As reported, the crystallinity index of cellulose nanocrystals is much higher compared to the starting materials i.e., cellulose microfibers. In the hydrolysis process, a highly crystalline CNCs material is obtained when the cellulose amorphous regions are removed while the highly ordered crystalline regions are kept intact. It is, however, in a certain case, nanocrystalline cellulose showed lower crystallinity index that is attributed to the source of nanocrystals i.e., isloated from recycled fibers. As reported from our published work [2], the crystallinity index of cellulose and nano crystalline cellulose from various sources are 54% and 90% respectively.

Current Research and Applications of Cellulose Nanocrystals

Composite Applications, Sensor, Super Capacitor, Solar Cell

Osamu Kose et al. 2019 [52] used cellulose nanocrystal in preparation of stretchable optics, Helbert et al. 1996 [27] had used CNC in fabrication of thermoplastic nanocomposites. Gaspar et al. 2014 [14] used CNCs in the preparation of composite with a gate dielectric properties. Favier et al. 1996 [53] used CNCs for preparation of polymer composite. To prepare a stretchable optics, Kose prepared glucose-cellulose nanocrystals film and then soaked it in an AIBN solution for 8 mins. Then, Elastomer and 2-HEA and AIBN were added. The result suggests that under certain stretched conditions and rotation the cellulose nanocrystal film exhibits birefringence [52]. Also, the CNC-E film stretches over 900%. Such composites are considered as a potential material for flexible optics and sensors. The author also suggested that CNC-E can be used as a versatile sensor. From Bras et al. 2010 [54], cellulose nanocrystals isolated from bagasse fibers using sulfuric acid treatment was combined with natural rubber at a loading from 0.0% up to 12.5% for composites fabrication. Then, those composites were investigated for water vapor permeability, thermal stability, biodegradation, and mechanical properties. The result shows that increase in loading of cellulose nanocrystal from 0% to 10% had significantly improved in the tensile strength and Young's modulus of natural rubber/cellulose composites. The thermal stability of natural rubber/cellulose whisker composites showed at par thermal stability compared to cellulose nanocrystals. In terms of barrier properties, a higher amount of water absorption with the reinforced cellulose whisker into the natural rubber matrix was observed. As reported, the water vapor permeability increases when the reinforced cellulose whiskers was incorporated into the matrix up to 7.5%, then any higher loading of cellulose whiskers decreased the water vapor permeabilites. Because of increase in density of cellulose network and formation of large enough 3D network, thus hydrogen bonding may help to reduce the water vapor permeability of composites. Yu et al. 2015 [23] used modified CNC to reinforce in silicone rubber matrix. CNC combined with IPTS that had enough isocyanate groups showed a better dispersion of modified CNCs within silicone rubber matrix. In addition, a high order of reinforcement effect which resulted in composites materials better tensile strength, elongation, tear strength and Young's modulus.

Cellulose nanocrystal aerogel are light weight which consists of high void volume has a potential use as a supercapacitor materials. In brief, the modification of cellulose nanocrystal to aldehyde- CNCs (CHO-CNCs) and hydrazide-modified CNCs ($NHNH_2$-CNCs) from APS produced COOH-CNCs. To prepare CNCs hydrogel, the mixture of PPy-NF, PPy-CNT or MnO_2-NP added to $NHNH_2$-CNC suspension and mixed with CHO-CNC. Then freeze-drying of suspension formed hybrid aerogels which is used for fabrication of supercapacitor cell, Yang et al. 2015 [55]. The aerogel showed good capacitor behavior, and excellent rate capability and retention of its ability at high scan rates up to 1000mV s^{-1}. Also reported, the application of CNC 3D aerogels substrates for supercapacitor materials. CNC 3D hydrogel is lightweight, strong compared to silica aerogels and can be deposited on top of a feather. The fabricated material has excellent capacitance retention at high charge-discharge rates and good cycle stability. Furthermore, such lightweight materials with good mechanical and capacitance indicator could be a potential candidate for large scale production of supercapacitors.

The fabrication route of the aerogel supercapacitor as follows:

$NHNH_2$-CNCs + CHO-CNCs + (PPy-NF or PPY-CNT or MnO_2-NO) → freeze ice-gel → freeze drying → Aerogel.

Petersson et al. 2007 [56], investigated the structure and thermal properties of CNC/PLA nanocomposites. Also, Lin et al. 2011 [57], reported cellulose nanocrystal surface modifications with acetylation and then reinforced into poly (lactic acid) matrix. At first, cellulose nanocrystal and pyridine were thoroughly dispersed using ultrasonic treatment for 15 mins. The surface modified CNCs were prepared by addition of CNC with acetic anhydride in anhydrous pyridine at 80°C, mixed with 400 rpm for 5 hours. At a certain conditions, the surface modified CNC was mixed with PLA. The reinforced component of modified CNC loading was from 1.0% to 10.0% in PLA matrix. By surface modification of CNCs, a better dispersion ability in different solvents and decrease surface polarity was observed. In addition, the tensile strength improved while adding from 1.0% to 6.0% of modified CNC into PLA after that the higher loading of modified CNC reduced the tensile strength of composites. However, the Young's modulus improved from 900 MPa reached to almost 1.30 GPa. Also, the authors indicated that the thermal stability of the composites improved by using surface modified CNCs. El Miri et al. 2016 [58] successful fabricated hybrid CNC/graphene oxide nanosheets and reinforced into PVA nanocomposites. The authors prepared different type of CNC composites that were mixed with a defined ratio of 5% CNC, graphene oxide, and CNC/Graphene oxide using a solvent casting method. The report indicates a significant improvement in tensile strength, Young's modulus, toughness and also the glass and melting temperatures. Vallejos et al. 2012 [16] prepared all-cellulose composite fibers that was formed by the electron spinning method with and without cellulose nanocrystal (~5.0%), and the composites with CNCs showed a decrease in water contact angle.

Solar Cell

Zhou et al. 2013 [59], Zhou et al. 2014 [60], reported cellulose nanocrystals as an ingredient for organic solar cells. From the authors, the CNC showed a good optical transparency and low surface roughness that is an important criteria's for prepartion of solar cells. Furthermore, the CNC used in solar cells are recyclable or combustion in a simple approach is possible. Costa et al. 2016 [61], used cellulose nanocrystal and cellulose nanofibrillated fiber in solar cells. The authors mentioned that the cellulose substrate was attached to the glass by double-sided tape and the solar cells configuration was Substrate/Ag/ZnO: Al/PFDTBTP: PC70BM/MoO_3/Ag. Guidetti et al. 2016 [62], prepared flexible photonic film with structured colors using mechansim of co-assembly that consists of Zwitterionic (Zw) surfactants with CNCs. Gaspar et al. 2014 [14], used CNC as a gate dielectric carrier and substrate in flexible field effect transistors.

Smart Materials

Dong and Roman [63], prepared smart materials by incorporating CNCs which is mainly used for measurement of fluorescence bioassay and bio imaging applications. For preparation, CNC was decorated with epoxy functional groups by reaction with epichlorohydrin with 1M NaOH. The reaction occurred at 60°C for 2 hours then followed by dialysis with deionized water to achieve pH less than 12. After that, the opened epoxy reaction occurred with ammonium hydroxide. Then, the sample was neutralized to achieve pH of 7.0. Finally, the amino group was reacted with the isothiocyanate group of FITC to form thiourea. Another research group, Nielsenet al. 2010 [64], used CNCs in the composite for assembly of pH sensor. According to the authors, CNC was combined with FITC and RBITC to form a thiocarbamate bond. During reaction, the fabricated CNC-FITC or CNC-RBITC will change the color from white to pink by which the pH can be detected. From Kim et al. 2015 [65], cellulose nanocrystal showed potential applications in thermo, pH, and magnetic responsive. Also, Hubber et al. 2008 [66], reported cellulose nanocrystal can be used in different applications such as drug delivery, tissue scaffolds, sensors and so on. Based on literature review, nanocrystal cellulose was reinforced into polymer matrix to produce shape memory materials which plays a significant role in intelligent/smart composites, or thereby grafting with poly (L-Lactic acid) with significant mechanical properties such as higher tensile strength and elongation at break [67-71].

Biomedical Applications

CNC used as a reinforcement agent in polyurethane produces shape memory materials that are mainly used for biomedical applications [68]. The results show that the reinforcement of CNCs as filler into polymer matrix improves the toughness, yield strength, and recovered rate of polyurethane composites that is crucial for biomedical applications. According to Gonzalez et al. 2014 [72], cellulose nanowhisker combined with poly (vinyl alcohol) forms nanocomposites hydrogels, a potential candidate for wound dressing applications. Such composite hydrogels can be prepared with varied concentrations of cellulose nanowhisker of 1, 3, 5 and 7% that is in corporated into PVA and continuously mixed for 3hrs. Then, the mixture was frozen for 60 mins and cooled at -18°C. As mentioned, the composites for wound dressing was fabricated with good mechanical and functional properties, as well as bacterial resistance. Authors also suggested that the nanocomposites materials with CNC loadings of 3.0% are considered as a promising materials for wound dressing. Yu et al. 2017 [35], reported use of cellulose nanocrystal/polyethylene glycol composite for controlling long term in-vitro drug release. A recent review from Du et al. 2019 [73] had reported applications of CNC and CNF for biomedical applications. For example, Prathapan et al. 2018 [74] used CNCs composite strips for the diagnosis of blood types. The diagnostics strip was prepared by PEO and CNC were coated on to PDMS using a spin coating method. Then, the PDMS coated PEO/CNC was transferred to PAE-coated glass substrate. In addition, the transparent nature of CNC substrate makes it is easy to conduct the cell analysis by optical imaging or spectroscopy.

Coating, Lubrication

From Satam et al. 2018 [75], cellulose nanocrystal and chitin nanofiber were coated on poly (lactic acid) film and investigated for barrier applications such as O_2 permeability and water vapor resistance. In brief, PLA as a fixed substrate and heated at 60°C during spray coating of CNCs and chitin nano fiber. The coating was conducted layer by layer deposition of chitin and cellulose nanocrystals on PLA substrate. The results show that coating with chitin and cellulose nanocrystals reduced the O_2 permeability compared to pure PLA. As noticed, a single layer coating of chitin nanofiber or cellulose nanocrystal showed no improvement for O_2 permeability. Authors also mentioned that the coating of chitin, cellulose nanocrystals or combination of those two materials is regarded a good bio-renewable barrier packaging material interms of oxygen permability and water vapor resistance. From Luzi et al. 2016 [76], cellulose nanocrystal was modified with acid phosphate ester of ethoxylated nonyl phenol to form (s-CNC) [72]. Also, the combination of CNC with PLA or PLA with a certain amount of ZnO were investigated for its effect on composite mechanical properties, thermal stability, antimicrobial, and barrier properties. The result shows an significant improvement intensile strength of the composite intercalated with CNCs, however, when combining PLA-CNC-ZnO the mechanical properties were decreased. In the case of barrier properties, water vapor permeability gained benefits. The authors suggest that the combination of PLA-1s-CNC-0.1ZnO is considered as a potential combination for functional packaging applications such as food wrappers and containers.

Automotive Sector

In general, cellulose nanofiber use in automotive interior parts have been extensively studied. The potential applications of cellulose fiber as a substitute to produce composite interior parts were explored by several automobiles' producers. Because of high production cost, the initial automotive composite parts production has been temporarily halted. In recent years, the issues of environmental impacts (recycle of waste), cost-effectiveness and the availability of new technology have led to reconsidering use of cellulose nanofiber for the automotive parts. Kiziltas et al. [77] also mentioned that around 30% of cellulose based nanomaterials can replace other materials such as glass fiber, talc and others. As indicated, the advantages of cellulose nanofiber properties such as its light weight and better mechanical properties and requiring less material in the interrior parts that significanlty reduces the total weight of an automobile and thus less energy consumption and resulting in potential cost savings. Furthermore, the large automotive firms have started using cellulose based composite materials as a substitute for their traditional products. There is no detailed information on applications of CNC in the automobile industry as reported by Yu et al. 2015 [23], Panchal et al. 2018 [78].

CNC Pilots Plant

Both industrial and pilot plant nanocellulose production facilities are located across the world. For ex., nanocellulose manufacturing facilities in Japan are owned by Nippon Paper group, Daio Paper, Chuetsu pulp and paper and others. It is, however, cellulose fibrils that are produced in those factories rather than cellulose nanocrystals. The pilot plants in Canada and the USA mainly produce CNCs (by acid hydrolysis) only. Celluforce owns the largest commerical scale CNC production facility which is located in Canada, the company produces around 300 tons of CNCs a year. Also, Celluforce has obtained about 20% of active registered patents in CNC production. Another player, InnoTech has major production facility in

Alberta that produces 100kgs of CNCs per week. Also, FPInnovations pilot scale plant produce CNCs. Another supplier of CNCs in USA, the University of Maine supports mostly academic, and private institution research studies. USDA/Forest product laboratory can produce 50 kgs of CNC per week. From Brazil, a pilot plant for CNC production was also setup with the ability to produce bio-based products.

Recently Published Patents on Cellulose Nanocrystal

As reported by Charreau et al. 2013 [79], the number of different types of nanocellulose (cellulose nanocrystal) related patent filed in early 1990's are quite a few patent applications. Since 2003, a steady growth in patent filing were observed (4-5) but an significant increase in CNC registered patent filing started from 2007 to 2011 (around 16 and 25). Table 3 shows the patent filings related to CNC and CNC applications that was limited to the WIPO website only. The current registered patent with searched titled "cellulose nanocrystals" in early June 2019 was 30 patents, 26 for 2018, and 50 for 2017. As of now, the result suggests an incredible increase in patents applications filings for CNC as illustrated in Table 3.

CONCLUSION

Cellulose nanocrystal is a enivronmental-friendly material, with abundant availability, and innumerable potential applications which has gained much of academic research interests and the industry attraction. From the search of patent applications, there is a strong and brighter future for cellulose nanocrystal use in different applications are evident and clear. However, the current issues are limited spread out information amongst wider audiences on eco-friendly CNCs and cellulose nanofiber products that are prepared from renewabe resources. Also, the ongoing cellulose nanofibers research development activities and installation of new small production facilities are limited by capital constraints. In order to propel the continued production of CNCs for commerical use and realizing the maximum potential benefits of using CNCs in various applications, the government funding support mechanism is must. The applications of cellulose nanocrystal are not only limited to pulp and paper products, but also for its prefential use in high-value products such as surface coating, lubricant, composite materials, cosmetics, food packaging, automotive interior parts, pharmaceutical products, and smart materials.

Table 3. Patent on CNC in WIPO from January to June 2019*

No	Publication number	Title	Inventors
1	20190169421	Cellulose nanocrystals-thermoset resin systems, applications thereof and articles made therefrom	Yaolin Zhang; Lamfeddal Kouisni; Xiang-Ming Wang; Michael Paleologou; Martin W. Feng; Gilles Brunette; Guangbo He;Hui Wan; Ayse Alemdam-thomson
2	20190127625	Modified Cellulose Nanocrystals and Their Use in Drilling Fluids	QinglinWu; Meichun Li
3	20190169797	Method of producing a carboxyalkylated NFC product, a carboxyalkylated NFC product and use thereof	Ali Naderi
4	WO/2019/105998	Double-layered cellulose nanofiber material, method of manufacturing, membranes, and use thereof	Mathew, Aji Pallikunnel; Liu, Peng
5	20190128828	Chemiresistor humidity sensor and fabrication method thereof	Sebran, Bogdan Catalin; Buiu, Octavian; Ionescu, Octavian; Buiu, Andrei;
6	WO/2019/082023	Composition containing a silicone-based adhesive and cellulose nanocrystals, and method and articles	Lipscomb, Corinne E.; Tse, Kiu-Yuen; Richardson, Jenna L.; Bénon, Karl E.
7	WO/2019/075184	Method to produce composite-enhanced market pulp and paper	Bilodeau, Michael A.; Paradis, Mark A.;
8	20190100604	Processes and apparatus for producing nanocellulose, and compositions and products produced therefrom	Kimberly Elson; Theodora Resita Vesa Pylkkanen; Ryan O'connor

No	Publication number	Title	Inventors
9	20190092926	Functionalized cellulose nanocrystal materials and methods of preparation	Jeffrey Paul; Youngblood; Youngman Yoo
10	20190093288	Controlled porosity structural material with nanocellulose fibers	Michael Darin Mason; David Gregg Holomakoff; Muhammad Radowan Hossen
11	WO/2019/058019	Nanocrystalline cellulose containing dental material	Sailynoja, Eija
12	WO/2019/058392	Shaped objects for use in security applications	Shanmuganathan, Kadhiravan; Venugopalan, Premnath; Ambone, Tushar
13	20190085511	Composition material comprising ultrafine cellulose fibers, and method for producing material comprising ultrafine cellulose fibers	Takayuki Shimaoka; Ikue Homma Moe Mizukami; Takuri Ozaki
14	201702572	Additive of chemically modified cellulose nanofibrils or cellulose nanocrystals	Lee J Hall; Jay P Deville Orlando J Rojas; Carlos A Carrilo Carlos L Salas
15	WO/2019/050819	Continuous roll-to-roll fabrication of cellulose nanocrystals (CNC) coatings	Youngblood; Jeffrey Paul; Chowdhury, Reaz; Nurddin, Md
16	20190062211	Cementitious inorganic material containing cellulosic nanofibers	Vivek Bindiganavile; Mehmet Yaman Boluk Mounir El-Bakkari; Jose Goncalves
17	WO/2019/036789	Matrices composed of biopolymers for lithium adsorption and enrichment	De Leao Rosenmann, Bernardo
18	3447085	Cellulose-containing resin composition and cellulosic ingredient	Miyoshi Takaaki; Yamasaki Naoaki; Ueno Koichi; Sada Takashi; Nagata Kazuya; Sanada Kazuaki
19	20190055373	Bacterial cellulose gels, process for producing and methods of use	Andrew J. Hess, Ivan I. Smalyukh Qingkun Liu, Joshua A. De La Cruz Blaise Fleury, Eldho Abraham, Vladyslav Cherpak, Bohdan Senyuk
20	20190048147	Nano cellulosic compositions	Michael L. Curry Donald White
21	WO/2019/032127	Mesoporous zeolites and methods for synthesis thereof	Pilyugina, Tatiana
22	3441436	Bright pigment dispersion and method for forming multilayer coating film	Itoh Masayuki; Narita Nobuhiko Kuramochi Tatsuo; Isaka Hisashi Okazaki Hirokazu
23	20190039054	Mesoporous zeolites and methods for the synthesis thereof	Tatiana Pilyugina

Table 3. (Continued)

No	Publication number	Title	Inventors
24	20190040158	Production crystalline cellulose	Sean Mcalpine; Jory Nakoneshny
25	WO/2019/026071	Antimicrobial coating material comprising nanocrystalline cellulose and magnesium cellulose and magnesium oxide and method of preparation thereof	Hanuka, Ezra; Zolkov, chen; Nevo, Yuval; Azeraff, Clarite
26	3433306	High internal phase emulsion foam having cellulose nanoparticles	Rowan Stuart J; Feke Donald L; Karimhani Vahid; Manas-Zloczower Ica; Zhao Boran; Hubrard Wade Monroe JR; Wingert Mazwell Joseph; Merrigan Steven Ray
27	20190023857	Method for Preparing Non-Acid-Treated Eco-Friendly Cellulose Nanocrystal, and Cellulose Nanocrystal Prepared Thereby	Jihoon Shin; Yeong Un Kim; Min Woo Lee
28	20190016642	Cellulose nanocrystal-modified ceramic blank and preparation method thereof	Tengfei Deng; Yanjuan Wang Ning Lin; Xiaohong Xu
29	WO/2019/011009	Method for preparing polypropylene carbonate/cellulose nanocrystal composite material	Xie, Xiaolin; Li, Xiaojing; Wang, Yong; Zhou, Xingping
30	20190002700	Cellulose-based organic pigments	Mark P. Andrews; Timothy Morse

ACKNOWLEDGMENT

This research was supported by National Research Foundation of Korea (NRF-2015R1A3A2066301).

REFERENCES

[1] Ummartyotin, S., Manuspiya, H. (2015). "A critical review on cellulose: From fundamental to an approach on sensor technology." *Renewable and Sustainable Energy Reviews,* 41, 402-412.

[2] Hai, L. V., Roy, S., Muthoka, R. M., Park, J. H., Kim, H. C., and Kim, J. (2019). "Re-Use and Recycling of Materials: Solid Waste Management and Water Treatment." *Waste Paper: A Potential Source for Cellulose Nanofiber and Bio-nanocomposite Applications.* 327-344.

[3] Moon, R. J., Martin, A., Nairn, J., Simonsen, J., and Youngblood, J., (2011). "Cellulose nanomaterials review: structure, properties and nanocomposites." *Chem Soc Rev.,* 40, 3941-3994.

[4] Dumanli, A. G., van der Kooij, H. M., Kamita, G., Reisner, E., Baumberg, J. J., Steiner, U., and Vignolini, S. (2016). "Digital Color in Cellulose Nanocrystal Films." *Applied materials & Interfaces,* 6, 12302-12306.

[5] Habibi, Y., Lucia, L., Rojas, O. J., (2010). "Cellulose nanocrystals: chemistry, self-assembly, and applications." *Chemical reviews,* 110, 3479-500.

[6] Picard, G., Simon, D., Kadiri, Y., LeBreux, J. D., Ghozayel, F., (2012). "Cellulose nanocrystal iridescence: a new model." *Langmuir: Journal of surfaces and colloids,* 28, 14799-807.

[7] Gray, D. G., and Mu, X. (2015). "Chiral Nematic Structure of Cellulose Nanocrystal Suspensions and Films; Polarized Light and Atomic Force Microscopy." *Materials,* 8, 7873-7888.

[8] Lee, Y. S. (2008) in "*Self-Assembly and Nanotechnology: A Force Balance Approach.*" John Wiley & Sons, Inc., 22-164.

[9] Jee, A. Y., Tsang, B, and Granick, S. (2015). "Colloidal phase transitions: A switch for phase shifting." *Nat. Mater.,* 14, 17.

[10] Hai, L. V., Seo, Y. B., and Narayanan, S. (2018). "Hardwood cellulose nanocrystals: multi-layered self-assembly with evident of circular and distinct nematic pitch." *Cellulose chemistry and technology,* 52, 597-601.

[11] Ranby, G., (1949). "Aqueous colloidal solutions of cellulose micelles." *Acta Chemica Scandinavica.,* 3, 649–50.

[12] Liu, D., Zhong, T., Chang, P. R., Li, K., Wu, Q., (2010). "Starch composites reinforced by bamboo cellulosic crystals." *Bioresource Technology,* 101, 2529-2536.

[13] Hai, L. V., Son, H. N., Seo, Y. B., (2015). "Physical and bio-composite properties of nanocrystallinecellulose from wood, cotton linters, cattail, and red algae." *Cellulose,* 22, 1789.

[14] Gaspar, D., Fernandes, S. N., de Oliveira, A. G., Fernandes, J. G., Grey, P., Pontes, R. V., Pereira, L., Martins, R., Godinho, M. H., and Fortunato, E., (2014). "Nanocrystalline cellulose applied simultaneously as the gate dielectric and the substrate in flexible field effect transistors." *Nanotechnology,* 25, 094008.

[15] El Miri, N., El Achaby, M., Fihri, A., Larzek, M., Zahouily, M., Abdelouahdi, K., Barakat, A., Solhy, A. (2016). "Synergistic effect of cellulose nanocrystals/graphene oxide nanosheets as functional hybrid nanofiller for enhancing properties of PVA nanocomposites." *Carbohydrate Polymers,* 137, 239-248.

[16] Vallejos, M. E., Peresin, M. S., Rojas, O. J., (2012). "All-Cellulose Composite Fibers Obtained by Electrospinning Dispersions of Cellulose Acetate and Cellulose Nanocrystals." *J polym Environ,* 20, 1075-1083.

[17] Seoane, I. T., Fortunati, E., Puglia, D., Cyras, V. P., and Manfredi, L. B., (2015). "Development and characterization of bionanocomposites based on poly(3-hydroxybutyrate) and cellulose nanocrystals for packaging applications." *Polym Int.,* 65, 1046-1053.

[18] Lei, W., Fang, C., Zhou, X., Yin, Q., Pan, S., Yang, R., Liu, D., Ouyang, Y., (2017). "Cellulose nanocrystals obtained from office waste paper and their potential application in PET packing materials." *Carbohydrate polymer,* 181, 376-385.

[19] Neto, F., Pires, W., Silvério, Alves, H., Dantas, Oliveira, N., Pasquini, Daniel (2013). "Extraction and characterization of cellulose nanocrystals from agro-industrial residue – Soy hulls." *Industrial Crops and Products,* 42, 480-488.

[20] Lu, P., Hsieh, Y. L., (2012). "Preparation and characterization of cellulose nanocrystals from rice straw." *Carbohydrate Polymers,* 1, 564- 573.

[21] Azizi Samir, M. A. S., Alloin, F., Sanchez, J. Y., and Dufresne, A. (2004). "Cross-Linked Nanocomposite Polymer Electrolytes Reinforced with Cellulose Whiskers." *Macromolecules* 37, 4839-4844.

[22] Pandey, J. K., Chu, W. S., Kim, C. S., Lee, C. S., Ahn, S. H. (2009). "Evaluation of morphological architecture of cellu- lose chains in grass during conversion from macro to nano dimensions." *E-polymers,* 1, 1221.

[23] Yu, H. Y., Chen, R., Chen, G. Y., Liu, L., Yang, X. G., Yao, J. M. (2015). "Silylation of cellulose nanocrystals and their reinforcement of commercial silicone rubber." *J Nano part Res,* 17, 361.

[24] Ranby, B. G. (1949). "Aqueous colloidal solutions of cellulose micelles." *Short communications*, 649 – 650.

[25] Maiti, S., Jayaramudu, J., Dasa, K., Reddy, S. M., Sadiku, R., Ray, S. S., Liu, D., (2013). "Preparation and characterization of nano-cellulose with new shape from different precursor." *Carbohydrate Polymers,* 98, 562– 567.

[26] Rosa, M. F., Medeiros, E. S., Malmonge, J. A., Gregorski, K. S., Wood, D. F., Mattoso, L. H. C., Glenn, G., Orts, W. J., Imam, S. H., (2010). "Cellulose nanowhiskers from coconut husk fibers: Effect of preparation conditions on their thermal and morphological behavior." *Carbohydrate Polymers,* 81, 83–92.

[27] Helbert, W., Cavaille, J. Y., and Dufresne, A., (1996). "Thermoplastic Nanocomposites Filled With Wheat Straw Cellulose Whiskers. Part I: Processing and Mechanical Behavior." *Polymer Composites,* 17.

[28] Edgar, C. D., & Gray, D. G., (2001). "Induced circular dichroism of chiral nematic cellulose films." *Cellulose,* 8, 5–12.

[29] Xua, Q., Gao, Y., Qin, M., Wu, K., Fu, Y., Zhao, J., (2013). "Nanocrystalline cellulose from aspen kraft pulp and its application in deinked pulp." *International Journal of Biological Macromolecules,* 60, 241– 247.

[30] Fahma, F., Iwamoto, S., Hori, N., Iwata, T., Takemura, A., (2010). "Isolation, preparation, and characterization of nanofibers from oil palm empty-fruit-bunch (OPEFB)." *Cellulose,* 17, 977–985.

[31] Kumar, A., Negi, Y. S., Choudhary, V., Bhardwaj, N. K., (2014). "Characterization of Cellulose Nanocrystals Produced by Acid-Hydrolysis from Sugarcane Bagasse as Agro-Waste." *Journal of Materials Physics and Chemistry,* 2, 1, 1-8.

[32] Danial, W. H., Majid, Z. A., Muhid, M. N. M., Triwahyono, S., Bakar, M. B., Ramli, Z. (2015). "The reuse of wastepaper for the extraction of cellulose nanocrystals." *Carbohydrate Polymers,* 118, 165-169.

[33] Kargarzadeh, H., Ahmad, I., Abdullah, I., Dufresne, A., Zainudin, S. Y., Sheltami, R. M., (2012). "Effects of hydrolysis conditions on the morphology, crystallinity, and thermal stability of cellulose nanocrystals extracted from kenaf bast fiber." *Cellulose,* 19, 855-866.

[34] Martins, D. F., de Souza, A. B., Henrique, M. A., Silvério, H. A., Neto, W. P. F., Pasquini, D., (2014). "The influence of the cellulose hydrolysis process on the structure of cellulose nanocrystals extracted from capim Mombasa (Panicum maximum)." *Industrial Crops and Products,* 65, 496-505.

[35] Yu, H. Y., Wang, C., Abdlkarim, S. Y. H., (2017). "Cellulose nanocrystals/polyethylene glycol as bifunctional reinforcing/ compatibilizing agents in poly(lactic acid)." *Cellulose,* 24, 4461-4477.

[36] dos Santos, R. M., Neto, W. P. F., Silvério, H. A., Martins, D. F., Dantas, N. O., Pasquini, D. (2013). "Cellulose nanocrystals from pineapple leaf, a new approach for the reuse of this agro-waste." *Industrial Crops and Products,* 50, 707-714.

[37] Camacho, M., Ureña, Y. R. C., Lopretti, M., Carballo, L. B., Moreno, G., Alfaro, B., and Baudrit, J. R. V., (2017). "Synthesis and Characterization of Nanocrystalline Cellulose Derived from Pineapple Peel Residues." *J. Renew. Mater.,* 5, 271- 279.

[38] Mohamad Haafiz, M. K., Hassan, A., Zakaria, Z., Inuwa, I. M., (2014). "Isolation and characterization of cellulose nanowhiskers from oil palm biomass microcrystalline cellulose." *Carbohydrate Polymers,* 103, 119-125.

[39] Yu, M., Yang, R., Huang, L., Cao, X., Yang, F., and Liu, D. (2012). "Preparation and characterization of Bamboo nanocrystalline cellulose." *Bioresources,* 7, 1802-1812.

[40] Luzi, F., Puglia, D., Sarasini, F., Tirillò, J., Maffei, G., Zuorro, A., Lavecchia, R., Kenny, J. M., Torre, L., (2019). "Valorization and Extraction of Cellulose Nanocrystals from North AfricanGrass: AmpelodesmosMauritanicus (Diss)." *Carbohydrate Polymers,* (accepted 2019).

[41] Johara, N., Ahmad, I., Dufresne, A., (2011). "Extraction, preparation and characterization of cellulose fibres and nanocrystals from rice husk." *Industrial crops and products*, 37, 93-99.

[42] Salam, A., Lucia, L. A., and Jameel, H. (2013). "A Novel Cellulose Nanocrystals-Based Approach to Improve the Mechanical Properties of Recycled Paper." *Sustainable chemistry & Engineering,* 1, 1584-1592.

[43] Cheng, M., Qin, Z., Chen, Y., Hu, S., Ren, Z., and Zhu, M., (2017). "Efficient Extraction of Cellulose Nanocrystals through Hydrochloric Acid Hydrolysis Catalyzed by Inorganic Chlorides under Hydrothermal Conditions." *Sustainable Chemistry & Engineering,* 5, 4656-4664.

[44] Yu, H., Qin, Z., Liang, B., Liu, N., Zhou, Z., and Chen, L., (2013). "Facile extraction of thermally stable cellulose nanocrystals with a high yield of 93% through hydrochloric acid hydrolysis under hydrothermal conditions." *Journal of materials chemistry,* 1, 3938.

[45] Jordan, H. J., Easson, M. W., Dien, B., Thompson, S., Condon, B. D. (2019). "Extraction and characterization of nanocellulose crystals from cotton gin motes and cotton gin waste." *Cellulose,* 1-21.

[46] Sadeghifar, H., Filpponen, I., Clarke, S. P., Brougham, D. F., Argyropoulos, D. S. (2011). "Production of cellulose nanocrystals using hydrobromic acid and click reactions on their surface." *J of Mater. Scien.*

[47] Torlopov, M. A., Udoratina, E. V., Martakov, I. S., Sitnikov, P. A. (2017). "Cellulose nanocrystals prepared in $H_3PW_{12}O_{40}$-acetic acid system." *Cellulose,* 24, 2153-2162.

[48] Liu, Y., Wang, H., Yu, G., Yu, Q., Li, B., Mu, X. (2014). "A novel approach for the preparation of nanocrystalline celluloseby using phosphotungstic acid." *Carbohydrate Polymers,* 110, 415-422.

[49] Luzi, F., Fortunati, E., Puglia, D., Lavorgna, Santulli, C., Kenny, J. M., Torre, L., (2014). "Optimized extraction of cellulose nanocrystals from pristine and carded hemp fibres." *Industrial Crops and Products,* 56, 175-186.

[50] Lu, Q., Cai, Z., Lin, F., Tang, L., Wang, S., and Huang, B. (2016). "Extraction of Cellulose Nanocrystals with a High Yield of 88% by Simultaneous Mechanochemical Activation and Phosphotungstic Acid Hydrolysis." *Sustainable chemistry & Engineering,* 4, 2156-2177.

[51] Garcıa-Garcıa, D., Balart, R., Lopez-Martinez, J., Ek, M., Moriana, R., (2018). "Optimizing the yield and physico-chemical properties of pine cone cellulose nanocrystals by different hydrolysis time." *Cellulose,* 25, 2925-2938.

[52] Kose, O., Tran, A., Lewis, L., Hamad, W. Y., & MacLachlan, M. J., (2019). "Unwinding a spiral of cellulose nanocrystals for stimuli-responsive stretchable optics." *Nature communications,* 10, 1-7.

[53] Favier, V., Chanzy, H., and Cavaille, J. Y. (1996). "Polymer Nanocomposites Reinforced by Cellulose Whiskers." *Macromolecules,* 28, 6365-6367.

[54] Bras, J., Hassan, M. L., Bruzesse, C., Hassan, E. A., El-Wakil, N. A., Dufresne, A. (2010). "Mechanical, barrier, and biodegradability properties of bagasse cellulose whiskers reinforced natural rubber nanocomposites." *Industrial Crops and Products,* 32, 627-633.

[55] Yang, X., Shi, K. Zhitomirsky, I., and Cranston, E. D. (2015). "Cellulose Nanocrystal Aerogels as Universal 3D Lightweight Substrates for Supercapacitor Materials." *Advanced materials,* 27, 6104-6109.

[56] Petersson, L., Kvien, I., Oksman, K., (2007). "Structure and thermal properties of poly(lactic acid)/cellulose whiskers nanocomposite materials." *Composites Science and Technology,* 67, 2535–2544.

[57] Lin, N., Huang, J., Chang, P. R., Feng, J., Yu, J. (2011). "Surface acetylation of cellulose nanocrystal and its reinforcing function in poly(lactic acid)." *Carbohydrate Polymers,* 83, 1834-1842.

[58] El Miri, N., Mounir, Achaby, E., Fihrib, A., Larzek, M., Zahouily, M., Abdelouahdi, K., Barakat, A., Solhy, A. (2016). "Synergistic effect of cellulose nanocrystals/graphene oxide nanosheets as functional hybrid nanofiller for enhancing properties of PVA nanocomposites." *Carbohydrate Polymers,* 137, 239-248.

[59] Zhou, Y., Fuentes-Hernandez, C., Khan, T. M., Liu, J. C., Hsu, J., Shim, J. W., Dindar, A., Youngblood, J. P., Moon, R. J., & Kippelen, B., (2013). "Recyclable organic solar cells on cellulose nanocrystal substrates." *Science reports,* 3, 1536.

[60] Zhou, Y., Khan, T. M., Liu, J. C., Fuentes-Hernandez, C., Shim, J. W., Najafabadi, E., Youngblood, J. P., Moon, R. J., Kippelen, B., (2014). "Efficient recyclable organic solar cells on cellulose nanocrystal substrates with a conducting polymer top electrode deposited by film-transfer lamination." *Organic electronics,* 15, 661-666.

[61] Costa, S. V., Pingel, P., Janietz, S., Nogueira, A. F. (2016). "Inverted organic solar cells using nanocellulose as substrate." *Journal of Applied polymer science,* 43679, 1-6.

[62] Guidetti, G. Atifi, S., Vignolini, S., and Hamad, W. Y. (2016). "Flexible Photonic Cellulose Nanocrystal Films." *Advanced materials,* 201603386.

[63] Dong, S., and Roman, M., (2007). "Fluorescently Labeled Cellulose Nanocrystals for Bioimaging Applications." *J. Am. Chem. Soc.*, 129, 13810–13811.

[64] Nielsen, L. J., Eyley, S., Thielemans, W., and Aylott, J. W., (2010). "Dual Fluorescent Labelling of Cellulose Nanocrystals for pH sensing." *Chemical Communications,* 46, 8929-8931.

[65] Kim, J. H., Shim, B. S., Kim, H. S., Lee, Y. J., Min, S. K., Jang, D., Abas, Z., and Kim, J., (2015). "Review of Nanocellulose for Sustainable Future Materials." *International journal of precision engineering and manufacturing-green technology,* 2, 197-213.

[66] Hubbe, M. A., Rojas, O. J., Lucia, L. A., and Sain, M. (2008). "Cellulosic nanocomposites: a review." *Bioresources,* 3, 929-980.

[67] Auad, M. L., Contos, V. S., Nutt, S., Aranguren, M. I., and Marcovich, N. E., (2008). "Characterization of nanocellulose reinforced shape memory polyurethanes." *Polym Int.*, 57:651–659.

[68] Auad, M. L., Mosiewicki, M. A., Richardson, T., Aranguren, M. I., Marcovich, N. E. (2009). "Nanocomposites Made from Cellulose Nanocrystals and Tailored Segmented Polyurethanes." *Journal of Applied Polymer Science,* 115, 1215–1225.

[69] Saralegi, A., Gonzalez, M. L., Valea, A., Eceiza, A., Corcuera, M. A. (2014). "The role of cellulose nanocrystals in the improvement of the shape-memory properties of castor oil-based segmented thermoplastic polyurethanes." *Composites Science and Technology,* 92, 27–33.

[70] Navarro-Baena, I., Kenny, J. M., Peponi, L. (2014). "Thermally-activated shape memory behaviour of bionanocomposites reinforced with cellulose nanocrystals." *Cellulose,* 21, 4231-4246.

[71] Garces, I. T., Aslanzadeh, S., Boluk, Y., and Ayranci, C., (2018). "Cellulose nanocrystals (CNC) reinforced shape memory polyurethane ribbons for future biomedical applications and design." *Journal of Thermoplastic Composite Materials,* 1-16.

[72] Gonzalez, J. S., Ludueña, L. N., Ponce, A., Alvarez, V. A., (2014). "Poly(vinyl alcohol)/cellulose nanowhiskers nanocomposite hydrogels for potential wound dressings." *Materials Science and Engineering C,* 34, 54-61.

[73] Du, H., Liu, W., Zhang, M., Si, C., Zhang, X., Li, B. (2019). "Cellulose nanocrystals and cellulose nanofibrils based hydrogels for biomedical applications." *Carbohydrate Polymers,* 209, 130-144.

[74] Prathapan, R., McLiesh, H., Garnier, G., and Tabor, R. F. (2018). "Surface engineering of transparent cellulosenanocrystal coatings for biomedical applications." *ASC applied bio materilas,* 1, 3, 728-737.

[75] Satam, C. C., Irvin, C. W., Lang, A. W., Jallorina, J. C. R., Shofner, M. L., Reynolds, J. R., and Meredith, J. C. (2018). "Spray-Coated Multilayer Cellulose Nanocrystal? Chitin Nanofiber Films for Barrier Applications." *Sustainable chemistry & engineering*, 6, 10637-10644.

[76] Luzi, F., Fortunati, E., Jiménez, A., Puglia, D., Chiralt, A., and Torre, L. (2016). "PLA Nanocomposites Reinforced with Cellulose Nanocrystals from Posidonia oceanica and ZnO Nanoparticles for Packaging Application." *Journal of Renewable Material,* 5, 103-115.

[77] Kiziltas, A., Kiziltas, E. E., Boran, S., & Gardner, D. J. (2013). "*Micro-and nanocellulose composites for automotive applications.*" http://www.temp.speautomotive.com/SPEA_CD/SPEA2013/pdf/BNF/BNF3.pdf.

[78] Panchal, P., Ogunsona, E., and Mekonnen, T. (2018). "Trends in Advanced Functional Material Applications of Nanocellulose." *Processes,* 7, 10.

[79] Charreau, H., Foresti, M. L., and Vázque, A. (2013). "Nanocellulose Patents Trends: A Comprehensive Review on Patents on Cellulose Nanocrystals, Microfibrillated and Bacterial Cellulose." *Recent patents on nanotechnology,* 7, 56-80.

In: Cellulose Nanocrystals
Editor: Orlene Croteau
ISBN: 978-1-53616-747-4

Chapter 3

CELLULOSE NANOCRYSTALS FROM BIORESOURCES AND THEIR APPLICATIONS

Sarthak Sharma[1] ***and Avnesh Kumari***[1,2,*]
[1]Nanobiology Lab, Biotechnology Division,
Council of Scientific and Industrial Research -
Institute of Himalayan Bioresource Technology (CSIR-IHBT),
Palampur (HP), India
[2]Academy of Scientific and Innovative Research,
New Delhi, India

ABSTRACT

Cellulose nanocrystals (CNCs) are exclusive nanomaterials (NMs) derived from the most abundant and almost inexhaustible natural polymer. These NMs have gathered the attention of the scientific community due to their unique mechanical, optical, chemical, and rheological properties. CNCs are biodegradable, biocompatible and renewable, hence serving as a sustainable and environmentally friendly material as they are mainly obtained from naturally occurring cellulose fibers. In view of the rising interdisciplinary research being carried out on CNCs, this review aims to

[*]Corresponding Author's Email: avnesh@ihbt.res.in; avneshv@yahoo.co.in.

assemble the knowledge available about the chemical structure, sources, physical and chemical procedures for the isolation of CNCs. The description of the mechanical, optical, and rheological properties of CNCs is also given in this review. Innovative applications in diverse fields such as pharmaceutics, catalysis, food, and packaging have also been discussed. In addition, latest advances in the fields of wound healing, regenerative medicines and drug delivery have also been highlighted.

1. INTRODUCTION

Cellulose is the most abundant organic polymer on the earth. It is the product of the biosynthesis of plants and animals and constitutes the main component of the cell wall in the plants (Moon et al. 2011). It is of great interest to the scientific community due to its tailored physical, chemical properties and biodegradable nature (Abitbol et al. 2016). In nature, cellulose is present in the form of bundles of individual cellulose forming fibres interconnected by hydrogen bonding. β-1,4-linked anhydro-D glucopyranose is a repeating monomer unit of cellulose which is also called as cellobiose (Wei et al. 2014). This monomer unit has a $^4C_1$ chair conformation and surfacial hydroxyl groups are positioned in the equatorial plane, while the hydrogen atoms are in the axial plane. Around 36 individual cellulose molecules packed together to form elementary photofibrils, which further combined to form microfibrils units resulted to form a cellulose fiber. Cellulose from different plant sources have the same basic chemical structure but differ physically due to different biosynthesis conditions, governed by the specific enzymatic terminal complexes, these terminal complexes give rise to crystalline and amorphous regions (Habibi et al. 2010). The amorphous region arises due to the chain dislocation at the micro fibril level and the crystalline region arises from the ordered fibrils inter connected by intra and inter molecular hydrogen bonding. Its ring-like structure provides high crystallanity. Cellulose inherent fascinating properties like biodegradability, biocompatibility, non-toxicity, higher water uptake capacity, higher mechanical strength, and abundance of hydroxyl groups lead to the active tunable surface area (Thomas et al. 2018). All these properties contribute to making cellulose as a self-reliant raw material for the applications in a number of fields like the pharmaceuticals industry, paper industry, textile, electronic sensors, composites industry and food industry (De France et al. 2017; Brinchi et al. 2013).

2. METHODS FOR ISOLATION OF CELLULOSE NANOCRYSTALS (CNCS)

Cellulose molecules with at least one of the dimensions in nanometers are termed as nanocellulose.

Nanocellulose can be divided into three types of materials:

1. Cellulose nanocrystals (CNCs), also known as nanocrystalline cellulose (NCC)
2. Cellulose nanofibrils (CNFs), also known as nano-fibrillated cellulose (NFC)
3. Bacterial cellulose (BC)

Amorphous regions of cellulose are cut down to obtain nanocrystalline cellulose or cellulose nanocrystals (CNCs). Moreover, breakdown of the cellulose to nanocellulose also increases the active surface area which is important for various surface modification related applications. The breakdown of cellulose is done by various mechanical, chemical and pretreatment methods optimized according to the dimensions of nanocellulose required (Saba et al. 2017). The pretreatment of the native cellulose is an efficient way of reducing the energy consumption of the fibrillation process. It could be achieved with approaches like enzyme hydrolysis and alkaline acid pretreatment. Enzyme pretreatment limits the use of harsh chemicals, hence leads to more economical and less toxic fibrillation. The mechanical fibrillation involves either chemical or enzymatic pretreatment methods to facilitate efficient fibrillation. Post treated fibers then processed through various mechanical processes like high-intensity ultrasonication, cryocrushing, homogenizing and microfluidization (Khalil et al. 2014; Nasir et al. 2017). On the other hand, the chemical process involves the hydrolysis in the presence of acids like HCl, H_2SO_4, H_3PO_4, HBr, and HNO_3. Acidic hydrolysis of cellulose gives us the nanocrystalline cellulose with high aspect ratios. Moreover, hydrolysis using H_2SO_4 and phosphoric acid introduces anionic groups on the surface of CNCs that leads to better stability and dispersion of CNCs in the aqueous medium. (2, 2, 6, 6-Tetramethylpiperidin-1-yl) oxyl radical aka TEMPO mediated oxidation is also a widely used chemical approach for the making of nanocellulose (Habibi et al. 2006). The unique properties of nanocellulose include high tensile strength, high water holding capacity, biocompatibility, chemical and physical modification. Different types of nanocellulose exhibit different characteristic properties depending on their dimensions, and source of origin.

2.1. Chemical Methods

CNCs are the product of acidic hydrolysis of native cellulose. Due to its higher mechanical strength and nano dimensions, it is being used in a wide range of applications in various fields. CNC is prepared by the acid hydrolysis of native cellulose. The choice of an acid depends on the application where we want to use it. H_2SO_4 (Xiang et al. 2003), HCl (Yu et al. 2013), H_3PO_4 (Gámez et al. 2006), HBr and HNO_3 has been extensively used for the hydrolysis of cellulose. It has been reported that on acidic hydrolysis using HCl the suspension of cellulose tends to flocculate and on high concentration, it degrades the cellulose (Araki et al. 1998). Moreover, the suspension formed after HCl hydrolysis showed non Newtonian behavior. On the other hand use of H_2SO_4 and phosphoric acid makes the suspensions not flocculating due to the introduction of charged groups on the surface of the cellulose. The density of charged surface groups was found lower in H_3PO_4 treated cellulose than in H_2SO_4 treated cellulose and negligible in the case of HCl. The presence of charged functional groups on the surface of H_2SO_4 treated cellulose helps to form stable dispersions but lacks thermal stability. Whereas H_3PO_4 treated cellulose are thermally stable even at high temperatures (van den Berg et al. 2007; Roman et al. 2004; Camarero Espinosa et al. 2013). The mechanical properties of CNCs were comparable in both cases. For the isolation of CNCs, the aqueous solution of raw cellulose is first subjected to acidic hydrolysis. Centrifugation is done, to remove unhydrolysed cellulose in the solution. Hydrolysed cellulose sample is then subjected to dialysis against double distilled water until the pH of the solution becomes neutral. One of the limitations of hydrolysis with H_2SO_4 or HCl is that it level off the degree of polymerization. Moreover the sulfate groups on the surface of CNCs are labile and can be easily removed under mild alkaline conditions. TEMPO mediated oxidation is another chemical process for making CNCs. It converts surface hydroxyl groups to stable and nonremovable carboxyl entities that lead to the better homogeneous dispersion of CNCs (Habibi et al. 2006).

2.2. Biological Methods

Cellulose is also prepared by biological methods. Many microorganisms produce polysaccharides such as cellulose which acts as a protective envelop around the cell. Cellulose produced by bacteria is popularly known as bacterial cellulose (BC). BC is synthesized as pure cellulose and it does not require pretreatment to remove lignin, pectin, and hemicelluloses before hydrolysis. It is produced by the bacteria Acetobacter xylinum (or Gluconacetobacter xylinus). It is the most efficient producer of BC (Lin et al. 2013). BC is secreted as a ribbon-shaped fibril, less than 100 - nm wide, which is composed of much finer 2-4 nm nanofibril. BC was first reported by Adrian Brown (1886) while working with Bacterium aceti in 1886. Astounding properties of BC like a high degree of crystallinity, tensile strength, water retention, and moldability make it interesting among scientific community. The synthesis of BC involves two mechanism viz., synthesis of uridine diphosphoglucose (UDPGIc) and then the polymerization of glucose into the micro fibrils. Synthesis of UDPGIc is done by the various chemical and biological processes such as gluconeogenesis, or the pentose phosphate cycle based on the carbon sources like pyruvate, glycerol, dicarboxylic acids, dihydroxyacetone hexoses. Static and agitated cultures are conventional ways to produce BC (Ross et al. 1991; Römling and Galperin 2015). In static cultures, medium is placed in a tray inoculated and cultured for 5-20 days until a cellulose membrane fills the tray. It results in the formation of dense gelatinous cellulose membrane. BC has high mechanical strength both in dry and wet state. It also shows high water absorbency and biocompatibility which makes it suitable for many potential applications.

3. Isolation of Cellulose Nanocrystals from Underutilised Bioresources

It is a well-known fact that cellulose is the most abundant biopolymer on earth. It is obtained from plant sources and certain microorganisms. In the current scenario, there is a rapid increase in demand for cellulose in various industries due to its biodegradable and non-toxic nature compared to other synthetic polymers. Taking in consideration its increasing demand and related environment concerns, the quest for the present research is to meet the increasing demand sustainably. In this context the use of underutilized bio resources and other biomass residue has been the topic of growing research. Using this approach of agro waste valorization, not only makes the production more economical but also reduces the environment concerns (dos et al. 2013). Agro waste including crop residues (Haafiz et al. 2013), rice husk (Johar et al. 2012), straw (Costa et al. 2013), sugarcane bagasse (Sun et al. 2004), garlic skin (Reddy & Rhim 2014), corncob residue (Liu et al.2016), tomato peels (Jiang et al. 2015), royal palm tree residue (Hafemann et al. 2019) and manure (Devi et al. 2015) are being used as a source of lignocelluloses and cellulose from which CNCs can be isolated using various physical and chemical methods These agro wastes are abundant and inexpensive; hence find their use in various industries

(Thomas et al. 2011). It is not only an efficient way for waste management but it also reduces the dependence on the other conventional sources

4. Applications of Cellulose Nanocrystals

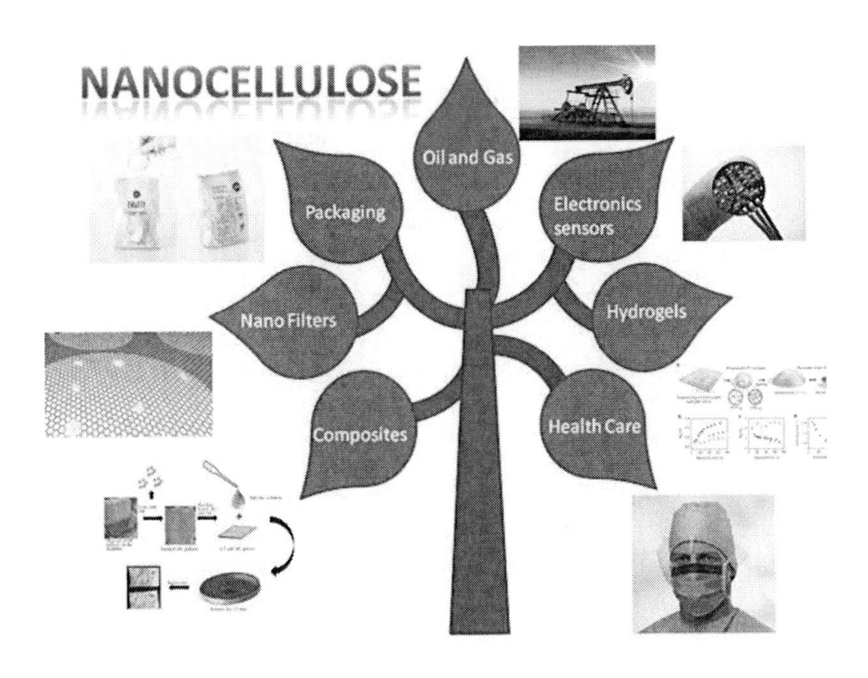

Figure 1. Applications of nanocellulose in different areas.

CNCs crystalline nature, tailorable physical, chemical properties and biodegradable nature make it compatible for many industries like construction, filtration, implants, hydrogel, packaging, paints, paper, pharmaceuticals, and electronics (Figure 1) (Brinchi et al. 2013). Due to the abundance of hydroxyl groups on the surface, there is an ease of surface modification. Its crystalline structure provides mechanical strength that makes it much stronger than steel or any other alloy. In composite industries, nanocellulose is being used as nanofiller in a variety of polymers to enhance its mechanical and thermal properties (Abitbol et al. 2016). Moreover, its biocompatibility and nontoxic nature makes it an essential biopolymer in regenerative medicine and tissue engineering. Its pH and temperature responsiveness make it an ideal carrier for controlled drug delivery (De France et al. 2017).

4.1. Applications in Pharmaceutical Industry

The use of nanocellulose is very wide in the field of pharmaceutical industries due to its fascinating properties like biocompatibility, biodegradability, higher mechanical strength, higher water holding capacity and ease of surface modification. CNCs find use as a drug excipient or drug delivery, immobilization and recognition of enzyme/protein as well as at the level of macroscopic biomaterials as blood vessel and soft tissue substitutes, skin and bone tissue repair materials, and antimicrobial materials (Lin & Dufresne, 2014). The surface fictionalization of cellulose plays a very vital role in enhancing the properties of cellulose and to obtain the desired results. Ethers and ether esters of cellulose are being produced by the functionalization of cellulose to obtain the polymers with enhanced hydrophobicity, increased thermoplasticity, high glass transition temperature, compressibility, and compatibility with drugs (Hale et al. 2018). Nanocellulose reinforced hydrogel with high water content has been used in making eye lenses and various other ophthalmic applications (Gopi et al. 2016). Microbial cellulose has also been used in various biomedical applications like the treatment of highly damaged skin and also as a scaffold for the regeneration of a variety of tissues (Wojciech et al. 2007). Nanocellulose based hydrogels have been used for making bio inks used in 3D bio printing of damaged soft tissues. 3D bioprinting have been used for soft tissues like nose, ear, meniscus, and cartilage in joints, as well as repair of damaged nerve tissue, and repair or replacement of damaged skin (Paul et al. 2018).

4.1.1. Wound Healing

Wound healing is the topic of great interest in the field of biomedical application. Wound repair is a complex coordinate process which involves various steps like coagulation and haemostasis, inflammation, proliferation and wound remodeling with scar tissue formation (Velnar et al. 2009). It has been reported that giving a moist environment to wound accelerates the re-epithilialisation and promotes faster tissue growth (winter, 1962). Keeping the wounds moist not only prevents the tissue dehydration, cell death but also increases the rate of angiogenesis; moreover, it reduces the sensation of pain by providing the cool sensation to the patients (Field & Kerstein 1994). This has led to the interest in producing an ideal wound dressing that accelerates the process of wound healing, reduce scarring and identification of signals that trigger the process of healing by regeneration rather than repair (Czaja et al. 2006). Nanocellulose hydrogels characteristic properties like ease to surface functionalization, higher water uptake, higher mechanical strength makes it an optimal material for wound dressings. Wound healing properties of nanocellulose has been evaluated *in vitro* and *in vivo* using Ca^{2+} crosslinked nanofibrillated cellulose hydrogel (Basu et al. 2018) and with silver nanoparticles (AgNPs) functionalized CNCs biocomposites (Singla et al. 2017). Both the studies concluded accelerated and efficient wound repair. Wood based nanofibrillated cellulose (NFC) have shown potential in wound healing in skin graft donor site treatment. Moreover, the NFC dressings were observed to adhere effectively to the wound bed and self-detaches after the wound recovery and reepithelialization. It was concluded that the rate of epithelialization or wound repair was faster in the case of NFC dressings as compared to the commercially available Suprathel® (Hakkarainen et al. 2016). BC produced by *Acetobacter xylinum* on other hand also served as an interesting wound dressing, due to its remarkable properties like higher water uptake, biocompatibility, ability to absorb exudates during the inflammatory phase and greater mechanical strength both in a dry and wet state (Fu et al. 2013). Some studies have shown the use of AgNPs onto BC for antimicrobial wound dressings. It was found that the silver ion functionalized BC exhibited strong antimicrobial activity against both *S. aureus* (Gram-positive bacteria) and *E. coli* (Gram-negative bacteria), (Maneerung et al. 2008). Another study has reported BC - chitosan (BC–Ch) as a wound dressing material which showed significant growth inhibition against *Escherichia coli* and *Staphylococcus aureus*. It has been reported that (BC–

Ch) showed better properties as a wound dressing material as compared to native BC and commercially available Tegaderm hydrocolloid or transparent films (Lin et al. 2013). Biocomposite hydrogel of bacterial nanocellulose and acrylic acid had been fabricated and used on to the wound site to study the behavior of human dermal fibroblasts (HDFs) at the cellular and molecular levels as a cell carrier and wound dressing material. The study concluded that the hydrogel is an efficient wound dressing material and HDF cell carrier to accelerate wound healing. At the cellular level, the hydrogel showed a rapid cell attachment and maintained the cell viability and morphology of HDFs and at the molecular level, the hydrogel unregulated eight important wound healing genes (IL6, IL10, MMP2, CTSK, FGF7, GM-CSF, TGFB1, and COX2) and downregulated one gene (F3) which could be beneficial during the wound-healing process (Xi Loh et al. 2018).

4.1.2. Regenerative Medicine

Regenerative medicine is an interdisciplinary field that deals with the recovery, repair and regrowth of the diseased or injured tissue or organs by using tissue engineering scaffolds to carry out the fabrication process of the infected tissue or organ. These biofabrications mimic the biological niche experienced by the cell and show better integration with the vasculature and nervous system of the host to create an environment where there is efficient regeneration of the tissue (Mao & Mooney 2015; Bailey 2014; Bajaj 2014). Tissue engineering scaffolds should have properties like biocompatibility and bioactivity, which further depends on the characteristic properties of the material like mechanical strength, topography, surface charge, chemical functionality, the aim is to provide a scaffold that mimics the microcellular environment of the infected tissues so that patients own cells migrate into the scaffold populate *in vivo* and results in to the regeneration of the lost tissue and *in vitro* the scaffolds are created in presence of therapeutic cells like autologous or allogeneic so that the material achieves the required biocompatibility for the transplantation on infected tissue (Place et al. 2009; Dugan et al. 2013). Biofabrication of the scaffolds was carried out using metals, ceramics and thermoplastic polymers that are processed using many organic solvents and other crosslinking agents, which makes them incompatible with the cellular matrix. Hydrogels provide a better favorable 3D aqueous environment for the encapsulation and growth for the cells similar to those in extra cellular matrix (Malda et al. 2013). Cell adhesion on the fabricated scaffolds has been the topic of research for many years. A study has been laid out showing the bio conjugation of two proteins, fibronectin and collagen type I on BC using 1-cyano-4-dimethylaminopyridinium (CDAP) tetrafluoroborate as catalyst. The CDAP activated scaffolds shown increased cell adhesion on the surface (Kuzmenko et al. 2013). Nanocellulose hydrogels due to its unique build-in properties serves as a potential material for the tissue scaffold and regenerative medicine (Thomas et al. 2018). Nanocellulose hydrogels have not shown any cytotoxicity. The only concern is its degradability in human body due to the absence of enzymes in human body that could break the glycosidic bonding and gives the monomeric glucose units, Studies have given a possible way of degradation by addition of enzymes, cellulases that could possibly degrade the cellulose and makes it bio-absorbable (Syverud 2017; Hu & Catchmark 2011). Nanocomposites hydrogels of Nanocellulose reinforced gellan gum hydrogel has been fabricated and taken into the use as biological substitutes for annulus fibrous tissue regeneration. Fabricated nanocomposites shown desired mechanical strength and significant cellular viability (Pereira et al. 2018). BC has been used as an artificial blood vessel in microsurgery (Klemm et al. 2001). BC and potato starch (PS) scaffolds have been used in hollow organ regeneration. Gelatinization properties of PS in presence of water have been incorporated with the BC to create a biomaterial with significant mechanical strength and pore size to ensure improved vessel formation and other hollow organ tissue engineering grafts (Lv et al. 2015). A similar study has shown the potential of bacterial nanocellulose as an implant material for ear cartilage replacement. Fabricated implants shape was patient specific and possess desired mechanical properties (Nimeskern et al. 2013). 3D Bioprinting is one of the latest advances in tissue engineering. Several studies have reported the use of nanocellulose as a bio ink for 3D printing of hydrogels. Another study has reported the use of nanocellulose incorporated with alginate to produce bio inks used in bio printing Human Chondrocytes. The 3D hydrogels produced were stable and compatible with the cells. Moreover, the mechanical strength of the scaffolds could be altered by changing the concentration of alginate depending on the application (Markstedt et al. 2015). Wood fiber derived acetylated nanocellulose as a bio ink in 3D scaffold showed significant compatibility with the myoblast cells. The scaffolds were stable, porous and showed better cell proliferation. Moreover, there is an absence of any cross-linking agent which makes it a single component bio ink and nontoxic. The research concluded that the scaffolds produced were cost efficient and could be produced fast (Ajdary et al. 2019). Nanocellulose due to their unique inherent properties and rheology shows potential in biomedicine as regenerative medicine and tissue engineering. However, the biocompatibility and cytocompatibility tests should be a mandate both *in vivo* and *in vitro* while using nanocellulose as biomedicine (Chinga-Carrasco 2018).

4.1.3. Drug Delivery

Conventional drug molecules use in pharma industries is limited due to poor solubility, tissue damage on extravasations, rapid breakdown of the drug, unfavorable pharmacokinetics, poor biodistribution, and lack of selectivity for target tissues. So there lies a dire need for an ideal drug delivery system that could enhance the therapeutic effect by a single-dose treatment that focuses on the sustained or prolonged release of the drug and targeting the active drug concentration to the site of action for the controlled release. This leads to an effective drug delivery with efficacy and flexibility. Efficient drug delivery system has many benefits (1) It decreases the amount of drug required (2) Due to the targeted delivery it significantly reduces the associated side effects (3) It reduces the number of doses to be taken by the patient hence increasing the patient compliance (4) It maintains the concentration of the active drug at the site of action all these factors contributes for the therapeutic effects (Deshpande et al. 1996; Langer 1998; Théron et al. 2014; Zhao et al. 2015). Many synthetic polymers (Uhrich et al. 1999), bio polysaccharides (Singh et al. 2011) have been used as carrier for slow and sustained release of drug. Among natural polymers, nanocellulose have also been used for drug delivery. Properties like biocompatibility, renewability and low toxicity makes them promising and safe alternative compared to synthetic polymers for controlled drug delivery. Cryogel composites of nanocellulose and gelatin have been used for slow and sustained release of anticancer drug 5-fluorouracil. The study concluded that 5-fluorouracil was encapsulated and successfully bonded to the composite by hydrogen bonding and both its release rate and cumulative release decreased with an increase in nanocellulose content, dialdehyde starch content, and density of the composite (Li et al. 2019). Another study has reported the use of surface modified CNCs with chitosan oligosaccharide as potential drug carriers for transdermal delivery applications. The drug carrier was formulated by the surface modification of TEMPO oxidized nanocellulose with chitosan oligosaccharide using N-hydroxysuccinimide and 1-ethyl-3-(3-dimethylaminopropyl)-carbodiimide as coupling agents. The study concluded that the modified nanocellulose particles showed a faster rate of drug release in 1 h at pH 8 and showed potential as a fast response drug carrier in local drug delivery (Akhlaghi et al. 2013). Another study reported use of polyethyleneimine-grafted cellulose nanofibres aerogels for drug delivery. The study concluded that the aerogels showed efficient sustained and controlled release which was dependent on pH and temperature (Zhao et al. 2015).

BC also find applications as carriers in drug delivery. Several studies have been reported showing the possibilities of BC as an efficient material for drug delivery applications. BC drug coating and drug release properties of the BC have also been investigated. With the number of remarkable properties like non toxicity, biodegradability and biocompatibility BC stand out as a promising material for the coating of drug tablets for sustained delivery. The BC films were thermally stable and comparable to commercially available Aquacoat *ECD* in terms of strength (Amin et al. 2012). Another study had shown the formulation of hydrogels of acrylic acid grafted on BC by exposure to accelerated electron-beam irradiation at different doses and its use in the controlled delivery system. The hydrogels showed dependence on pH and temperature and have been used for controlled delivery of protein based drugs. Hydrogel of BC- carboxymethyl cellulose (Pavaloiu et al. 2014) have been used for encapsulating Ibuprofen sodium salt. The hydrogel showed controlled delivery and swelling depended on the concentration of the carboxymethyl cellulose. At last, we can say that efficient drug delivery is a reliable approach for better therapeutic effects with less toxicity involved and ensuring patients' compliance.

4.2. Food Industry

Due to properties like biocompatibility, biodegradability, chemical stability and nontoxic nature, nanocellulose serves as a promising material for use in food industry as a food additive and packaging material. Cellulose was first time introduced in the late 80s as a food additive by Turbak and co-workers. They reported cellulose as a suspending medium for other solids and an emulsifying base for organic liquids (Turbak et al. 1983). However, it was not then marketed due to the issues related to energy consumptions and higher production costs. Several studies have reported the use of nanocellulose for enhancing the homogeneity and stability in a wide variety of food items and many patents have been filed indicating the use of nanocellulose as a food additive and a food packaging material. Moreover, nanocellulose has also been reported as a low calorie dietary fiber food additive used to lower the energy of processed food by holding the significant moisture content (Robson 2012; Serpa et al. 2016). A study has been reported showing the usage of cellulose as a food additive in hamburger. The study concluded that cellulose shows potential as a stabilizing agent for oil in water emulsions and foams without changing the food texture (Ström et al. 2013). Use of BC has also been reported as a low cholesterol diet due to its higher water uptake cation-exchange capabilities and lowering cholesterol properties (Chau et al. 2008). A study has been done showing the formulation for vegetarian meat using BC achieved by fermentation of BC using *Monascus purpureus* showing potential as a new foodstuff (Sheu et al. 2000). If we talk about food packaging, the current trend of food packaging lies in synthetic petroleum based polymers that are not only harmful to us but also for the nature. Several studies have been done in this direction in order to develop an ecofriendly biodegradable packaging material possessing significant physical strength and barrier properties (Khalil et al. 2016). Nanocellulose being biocompatible and biodegradable serves as a sustainable material for the production of bio composite membranes and films that possess significant mechanical strength and barrier properties. Crystalline structure, high surface area, and tailorable surface chemistry makes nanocellulose a potential barrier for the oxygen and water vapors. Several studies have been reported mentioning the barrier properties of cellulose films and coatings. A study has reported that using the coatings of microfibrillated cellulose at low humidity decreases oxygen permeability which is the same as that of synthetic polymers. The inter and intra hydrogen bonding of the mircofibrils were believed to be the contributing factor to this low permeability (Aulin et al. 2010). A similar study has reported the potential of nanobiocomposites formed using nanocellulose, acrylated epoxidized soybean oil and 3-aminopropyltriethoxysilane in low water vapor transmission rate (WVTR) (Lu et al. 2014). Formation of the nanocellulose composites and self-standing nanocellulose membranes could be achieved by various technological means like 1.Layer-by-layer assembly 2.Electrospinning 3.Composite extrusion 4.Coating 5.Casting from solution and evaporation. Several studies have been reported showing the potential use of nanocellulose as food packaging material. Many nanocellulose composites films are being made by grafting the synthetic polymers on to the CNCs by using grafting on to approach and taken into use for the food packaging material (Serpa et al. 2016; Dhar et al. 2014; Ferrer et al. 2017). Keeping in mind the requirements of ideal food packing new alternate biocomposites are being formulated that are not only better in barrier properties and mechanical strength but also play a significant role in increasing the shelf life of the food material. Several studies have been done using chitosan – nanocellulose films for extending shelf life. Chitosan, the derivative of chitin after deacetylation, is a bio polymer having a range of applications due to its inherent properties like biodegradability, biocompatibility, and antimicrobial activity. Many studies have shown the significance of chitosan based nanocellulose films in food packaging. Bio-nanocomposites produced were reported thermally stable as compared to synthetic films and shown an antimicrobial inhibition against gram-positive (*S. aureus*) and gram-negative bacteria (*E. coli* and *S. enteritidis*). Hence, they have shown potential as a better food packaging material for increasing the shelf life (Dehnad et al. 2014; Wang et al. 2018). Another study focused on the antimicrobial activity of nanocomposite films formed by the dispersion of nanocellulose in poly lactic acid matrix with the functionalization of AgNPs. The study concluded the formation of films with higher mechanical strength, better transparency, higher thermal stability and bacterial resistance against *S. aureus* and *E. coli* (Fortunati et al. 2012). We can conclude from the above discussion that due to the rising concern for the synthetic polymers and food additives we need to focus on the sustainable and biodegradable materials. Nanocellulose with several remarkable properties like biocompatibility, biodegradability, no toxicity and abundance in nature makes it a potential material to be used in the food industry as a food additive and in packaging applications.

4.3. Catalysis

Catalysts play a vital role in research and industrial manufacturing to make it more economical and to speed up the production process. Many of them are synthesized by inorganic chemicals which makes them toxic, environment polluting and expensive to use. On the other hand, cellulose is a biopolymer with remarkable properties like non toxicity, biodegradability, reusability, impressing water wettability, ease of surface modification and its insolubility for common solvents and hence stands out as an amazing material for its use as a heterogeneous catalytic system. Many studies have shown the potential of nanocellulose as a catalyst for various chemical reactions. Tailorable surface chemistry and abundance of hydroxyl groups in nanocellulose provide an ease to the surface modifications that lead to the formation of various nanoparticle-nanocellulose hybrid composites like metal nanoparticle-nanocellulose composite that has been extensively used towards catalysis. These hybrids heterogeneous catalytic systems have the potential of being biodegradable and reusable (Osanlou et al. 2017). Several studies have been done using Metal NP-nanocellulose hybrid composites as a catalyst in reduction, oxidation, coupling reaction, electro catalysis, photocatalysis, enzyme mobilization and biosensing (Kaushik & Moores 2016). Another study has been reported showing the potential of AuNPs (Gold nanoparticles) decorated on crystalline cellulose nanofibres as a catalyst in the reduction of 4-nitrophenol (4-NP) to 4-aminophenol (4-AP) with $NaBH_4$ (Koga et al. 2010). Similar studies have been laid out showing the catalytic activity of Ag (An et al. 2017), Pd (Wu et al. 2013) and Cu (Bendi & Imae et al. 2013) nanoparticle-nanocellulose hybrid composites in the reduction processes. RuNPs (Ruthenium nanoparticles)-nanocellulose composite hybrids, on the other hand, had shown a catalytic activity for the oxidation of benzyl alcohol to benzaldehyde (Gopiraman et al. 2015). Cobalt (II) supported on ethylenediamine-functionalized nanocellulose have been documented as potential catalyst for aerobic oxidation of alcohols at room temperature with good stability and recyclability (Shaabani et al. 2014). A study has been laid out showing the formation of robust, reusable sponges of cellulose nanofibres by dual crosslinking cellulose nanofibres with γ-glycidoxypropyltrimethoxysilane and polydopamine functionalized with PdNPs (Palladium nanoparticles) which was then used as catalysis in Suzuki and Heck cross-coupling reactions. These sponges were reusable and showed negligible leaching. It can also support different metal NPs catalysis like Ag, Au, and PtNPs (Platinum nanoparticles) and catalyze many other reactions like reduction of methylene blue and 4-nitrophenol, palladium-catalyzed C-H arylation, Sonogashira reactions reported in the study (Li et al. 2017). A Similar study has been carried out using CuNPs (Copper nanoparticles) functionalized nanocellulose as heterogeneous catalyst systems for C-N coupling reactions at room temperature (Goswami & Das 2018). A study has shown the potential of cellulose nanofibres as a hydrolytic catalyst with chiral specificities in the hydrolysis of enantiomeric amino acid substrates. The study concluded that the hydrolysis of amide bonds of amino acids was specific of chirality and can be classified as artificial enzymes (Serizawa et al. 2013). 2-methoxy-4-methylphenol

(MMP) utilized PdNPs decorated cellulose as a catalyst for its hydrodeoxygenation from vanillin. Conversion of vanillin to MMP has shown a remarkable activity and selectivity under mild conditions. Moreover, the catalyst can be easily separated and reused for 4 cycles with no significant loss in activity (Li et al. 2018). At last, we can conclude that due to the remarkable properties like tunable surface, non-toxicity, biodegradability, and abundance of this bio polymer in nature gives nanocellulose a potential to serve as a recyclable heterogeneous catalytic system in several reaction.

Conclusion

In this article, we have discussed the fascinating properties and pliant nature of nanocellulose, its application and the latest advances in the fields of regenerative medicine, wound healing, drug delivery, catalysis and in food and packaging industry. Isolation of nanocellulose is done by various physical, mechanical, and chemical processes. Pretreatment is required before mechanical disintegration as it makes the process economical. Wherein, chemical process acids like H_2SO_4, HCl, HBr, and H_3PO_4 are being used. The choice of the acid depends on the dimensions, surface charge and degree of polymerization of CNCs required. Acid hydrolysis of cellulose mainly depends on the acid concentration, temperature, and duration of the reaction. Nanocellulose is being widely used in pharmaceutical due to its remarkable inherent properties like biodegradability, biocompatibility, higher water uptake, non-toxicity, and tunable surface chemistry. Hydrogels and composites of nanocellulose are being extensively used for wound healing, regenerative medicine, and other drug delivery systems as they are capable of building efficient scaffolds and better encapsulation of the drug in the extra cellular matrix, which leads to an effective therapeutic result. Whereas, due to the properties like non-toxicity, biodegradability and great barrier properties nanocellulose is used as a food additive in low cholesterol diet and food packaging material in the food industry. Nanocellulose has also well-known uses in heterogeneous catalytic systems due to its biodegradable, recyclable and tunable surface chemistry. At last, we can say that nanocellulose is serving as an efficient material in various industries for many future applications.

Acknowledgments

The authors are thankful to the Director, CSIR-IHBT for his constant support and encouragement. SS acknowledges the financial support from CSIR, GOI in the form of MLP0141.

References

Abitbol, Tiffany, Amit Rivkin, Yifeng Cao, Yuval Nevo, Eldho Abraham, Tal Ben-Shalom, Shaul Lapidot, and Oded Shoseyov. "Nanocellulose, a tiny fiber with huge applications." *Current opinion in biotechnology* 39 (2016): 76-88.

Ajdary, Rubina, Siqi Huan, Nazanin Zanjanizadeh Ezazi, Wenchao Xiang, Rafael Grande, Hélder A. Santos, and Orlando J. Rojas. "Acetylated nanocellulose for single-component bioinks and cell proliferation on 3D-printed scaffolds." *Biomacromolecules* (2019).

Akhlaghi, Seyedeh Parinaz, Richard C. Berry, and Kam C. Tam. "Surface modification of cellulose nanocrystal with chitosan oligosaccharide for drug delivery applications." *Cellulose* 20, no. 4 (2013): 1747-1764.

Amin, M. C. I. M., Abadi Gumah Abadi, Naveed Ahmad, Haliza Katas, and Jamia Azdina Jamal. "Bacterial cellulose film coating as drug delivery system: physicochemical, thermal and drug release properties." *Sains Malaysiana* 41, no. 5 (2012): 561-568.

An, Xingye, Yunduo Long, and Yonghao Ni. "Cellulose nanocrystal/ hexadecyltrimethylammonium bromide/silver nanoparticle composite as a catalyst for reduction of 4-nitrophenol." *Carbohydrate polymers* 156 (2017): 253-258.

Araki, Jun, Masahisa Wada, Shigenori Kuga, and Takeshi Okano. "Flow properties of microcrystalline cellulose suspension prepared by acid treatment of native cellulose." *Colloids and Surfaces A: Physicochemical and Engineering Aspects* 142, no. 1 (1998): 75-82.

Arca, Hale Cigdem, Laura I. Mosquera-Giraldo, Vivian Bi, Daiqiang Xu, Lynne S. Taylor, and Kevin J. Edgar. "Pharmaceutical applications of cellulose ethers and cellulose ether esters." *Biomacromolecules* 19, no. 7 (2018): 2351-2376.

Aulin, Christian, Mikael Gällstedt, and Tom Lindström. "Oxygen and oil barrier properties of microfibrillated cellulose films and coatings." *Cellulose* 17, no. 3 (2010): 559-574.

Bailey, Alexander M., Michael Mendicino, and Patrick Au. "An FDA perspective on preclinical development of cell-based regenerative medicine products." *Nature biotechnology* 32, no. 8 (2014): 721.

Bajaj, Piyush, Ryan M. Schweller, Ali Khademhosseini, Jennifer L. West, and Rashid Bashir. "3D biofabrication strategies for tissue engineering and regenerative medicine." *Annual review of biomedical engineering* 16 (2014): 247-276.

Basu, Alex, Gunta Celma, Maria Strømme, and Natalia Ferraz. "*In vitro* and *in vivo* evaluation of the wound healing properties of nanofibrillated cellulose hydrogels." *ACS Applied Bio Materials* 1, no. 6 (2018): 1853-1863.

Bendi, Ramaraju, and Toyoko Imae. "Renewable catalyst with Cu nanoparticles embedded into cellulose nano-fiber film." *RSC Advances* 3, no. 37 (2013): 16279-16282.

Brinchi, L., F. Cotana, E. Fortunati, and J. M. Kenny. "Production of nanocrystalline cellulose from lignocellulosic biomass: technology and applications." *Carbohydrate Polymers* 94, no. 1 (2013): 154-16.

Camarero Espinosa, Sandra, Tobias Kuhnt, E. Johan Foster, and Christoph Weder. "Isolation of thermally stable cellulose nanocrystals by phosphoric acid hydrolysis." *Biomacromolecules* 14, no. 4 (2013): 1223-1230.

Chau, Chi-Fai, Pat Yang, Chao-Ming Yu, and Gow-Chin Yen. "Investigation on the lipid-and cholesterol-lowering abilities of biocellulose." *Journal of agricultural and food chemistry* 56, no. 6 (2008): 2291-2295.

Chinga-Carrasco, Gary. "Potential and limitations of nanocelluloses as components in biocomposite inks for three-dimensional bioprinting and for biomedical devices." *Biomacromolecules* 19, no. 3 (2018): 701-711.

Costa, Sirlene M., Priscila G. Mazzola, Juliana CAR Silva, Richard Pahl, Adalberto Pessoa Jr, and Silgia A. Costa. "Use of sugar cane straw as a source of cellulose for textile fiber production." *Industrial Crops and Products* 42 (2013): 189-194.

Czaja, Wojciech K., David J. Young, Marek Kawecki, and R. Malcolm Brown. "The future prospects of microbial cellulose in biomedical applications." *Biomacromolecules* 8, no. 1 (2007): 1-12.

Czaja, Wojciech, Alina Krystynowicz, Stanislaw Bielecki, and R. Malcolm Brown Jr. "Microbial cellulose—the natural power to heal wounds." *Biomaterials* 27, no. 2 (2006): 145-151.

De France, Kevin J., Todd Hoare, and Emily D. Cranston. "Review of hydrogels and aerogels containing nanocellulose." *Chemistry of Materials* 29, no. 11 (2017): 4609-4631.

Dehnad, D., Mirzaei, H., Emam-Djomeh, Z., Jafari, S. M., & Dadashi, S. (2014). Thermal and antimicrobial properties of chitosan–nanocellulose films for extending shelf life of ground meat. *Carbohydrate polymers, 109*, 148-154.

Deshpande, A. A., C. T. Rhodes, N. H. Shah, and A. W. Malick. "Controlled-release drug delivery systems for prolonged gastric residence: an overview." *Drug Development and Industrial Pharmacy* 22, no. 6 (1996): 531-539.

Devi, Rashmi R., Prodyut Dhar, Ajay Kalamdhad, and Vimal Katiyar. "Fabrication of cellulose nanocrystals from agricultural compost." *Compost science & utilization* 23, no. 2 (2015): 104-116.

Dhar, Prodyut, Umesh Bhardwaj, Amit Kumar, and Vimal Katiyar. "Cellulose nanocrystals: a potential nanofiller for food packaging applications." *Food additives and packaging* 1162 (2014): 197-239.

Dos Santos, Roni Marcos, Wilson Pires Flauzino Neto, Hudson Alves Silvério, Douglas Ferreira Martins, Noélio Oliveira Dantas, and Daniel

Pasquini. "Cellulose nanocrystals from pineapple leaf, a new approach for the reuse of this agro-waste." *Industrial Crops and Products* 50 (2013): 707-714.

Dugan, James M., Julie E. Gough, and Stephen J. Eichhorn. "Bacterial cellulose scaffolds and cellulose nanowhiskers for tissue engineering." *Nanomedicine* 8, no. 2 (2013): 287-298.

Ferrer, Ana, Lokendra Pal, and Martin Hubbe. "Nanocellulose in packaging: Advances in barrier layer technologies." *Industrial Crops and Products* 95 (2017): 574-582.

Field, Charles K., and Morris D. Kerstein. "Overview of wound healing in a moist environment." *The American journal of surgery* 167, no. 1 (1994): S2-S6.

Fortunati, E., I. Armentano, Qi Zhou, A. Iannoni, E. Saino, L. Visai, Lars A. Berglund, and J. M. Kenny. "Multifunctional bionanocomposite films of poly (lactic acid), cellulose nanocrystals and silver nanoparticles." *Carbohydrate polymers* 87, no. 2 (2012): 1596-1605.

Fu, Lina, Jin Zhang, and Guang Yang. "Present status and applications of bacterial cellulose-based materials for skin tissue repair." *Carbohydrate polymers* 92, no. 2 (2013): 1432-1442.

Gámez, Sara, Juan Jose González-Cabriales, José Alberto Ramírez, Gil Garrote, and Manuel Vázquez. "Study of the hydrolysis of sugar cane bagasse using phosphoric acid." *Journal of food engineering* 74, no. 1 (2006): 78-88.

Gatenholm, Paul, Hector Martinez, Erdem Karabulut, Matteo Amoroso, Lars Kölby, Kajsa Markstedt, Erik Gatenholm, and Ida Henriksson. "Development of nanocellulose-based bioinks for 3D bioprinting of soft tissue." *3D Printing and Biofabrication* (2018): 331-352.

Gopiraman, Mayakrishnan, Hyunsik Bang, Guohao Yuan, Chuan Yin, Kyung-Hun Song, Jung Soon Lee, Ill Min Chung, Ramasamy Karvembu, and Ick Soo Kim. "Noble metal/functionalized cellulose nanofiber composites for catalytic applications." *Carbohydrate polymers* 132 (2015): 554-564.

Goswami, Monmi, and Archana Moni Das. "Synthesis of cellulose impregnated copper nanoparticles as an efficient heterogeneous catalyst for CN coupling reactions under mild conditions." *Carbohydrate polymers* 195 (2018): 189-198.

Haafiz, MK Mohamad, S. J. Eichhorn, Azman Hassan, and M. Jawaid. "Isolation and characterization of microcrystalline cellulose from oil palm biomass residue." *Carbohydrate polymers* 93, no. 2 (2013): 628-634.

Habibi, Youssef, Henri Chanzy, and Michel R. Vignon. "TEMPO-mediated surface oxidation of cellulose whiskers." *Cellulose* 13, no. 6 (2006): 679-687.

Habibi, Youssef, Lucian A. Lucia, and Orlando J. Rojas. "Cellulose nanocrystals: chemistry, self-assembly, and applications." *Chemical reviews* 110, no. 6 (2010): 3479-3500.

Hafemann, Eduardo, Rodrigo Battisti, Cintia Marangoni, and Ricardo AF Machado. "Valorization of royal palm tree agroindustrial waste by isolating cellulose nanocrystals." *Carbohydrate Polymers* 218 (2019): 188-198.

Hakkarainen, T., R. Koivuniemi, M. Kosonen, C. Escobedo-Lucea, A. Sanz-Garcia, J. Vuola, J. Valtonen et al. "Nanofibrillar cellulose wound dressing in skin graft donor site treatment." *Journal of controlled release* 244 (2016): 292-301.

Hu, Yang, and Jeffrey M. Catchmark. "*In vitro* biodegradability and mechanical properties of bioabsorbable bacterial cellulose incorporating cellulases." *Acta biomaterialia* 7, no. 7 (2011): 2835-2845.

Jiang, Feng, and You-Lo Hsieh. "Cellulose nanocrystal isolation from tomato peels and assembled nanofibers." *Carbohydrate Polymers* 122 (2015): 60-68.

Johar, Nurain, Ishak Ahmad, and Alain Dufresne. "Extraction, preparation and characterization of cellulose fibres and nanocrystals from rice husk." *Industrial Crops and Products* 37, no. 1 (2012): 93-99.

Kaushik, Madhu, and Audrey Moores. "nanocelluloses as versatile supports for metal nanoparticles and their applications in catalysis." *Green Chemistry* 18, no. 3 (2016): 622-637.

Khalil, HPS Abdul, Y. Davoudpour, Chaturbhuj K. Saurabh, Md S. Hossain, A. S. Adnan, R. Dungani, M. T. Paridah et al. "A review on nanocellulosic fibres as new material for sustainable packaging: Process and applications." *Renewable and Sustainable Energy Reviews* 64 (2016): 823-836.

Khalil, HPS Abdul, Y. Davoudpour, Md Nazrul Islam, Asniza Mustapha, K. Sudesh, Rudi Dungani, and M. Jawaid. "Production and modification of nanofibrillated cellulose using various mechanical processes: a review." *Carbohydrate polymers* 99 (2014): 649-665.

Klemm, Dieter, Dieter Schumann, Ulrike Udhardt, and Silvia Marsch. "Bacterial synthesized cellulose—artificial blood vessels for microsurgery." *Progress in polymer science* 26, no. 9 (2001): 1561-1603.

Koga, Hirotaka, Eriko Tokunaga, Mami Hidaka, Yuuka Umemura, Tsuguyuki Saito, Akira Isogai, and Takuya Kitaoka. "Topochemical synthesis and catalysis of metal nanoparticles exposed on crystalline cellulose nanofibers." *Chemical communications* 46, no. 45 (2010): 8567-8569.

Kuzmenko, Volodymyr, Sanna Sämfors, Daniel Hägg, and Paul Gatenholm. "Universal method for protein bioconjugation with nanocellulose scaffolds for increased cell adhesion." *Materials Science and Engineering: C* 33, no. 8 (2013): 4599-4607.

Langer, Robert. "Drug delivery and targeting." *Nature-London-* (1998): 5-10.

Li, Dan-dan, Jia-wei Zhang, and Chun Cai. "Pd nanoparticles Supported on Cellulose as a catalyst for vanillin conversion in aqueous media." *The Journal of organic chemistry* 83, no. 14 (2018): 7534-7538.

Li, Jian, Yujia Wang, Lei Zhang, Zhaoyang Xu, Hongqi Dai, and Weibing Wu. "Nanocellulose/gelatin composite cryogels for controlled drug release." *ACS Sustainable Chemistry & Engineering* (2019).

Li, Yingzhan, Lei Xu, Bo Xu, Zhiping Mao, Hong Xu, Yi Zhong, Linping Zhang, Bijia Wang, and Xiaofeng Sui. "Cellulose sponge supported palladium nanoparticles as recyclable cross-coupling catalysts." *ACS applied materials & interfaces* 9, no. 20 (2017): 17155-17162.

Lin, Ning, and Alain Dufresne. "Nanocellulose in biomedicine: Current status and future prospect." *European Polymer Journal* 59 (2014): 302-325.

Lin, Shin-Ping, Iris Loira Calvar, Jeffrey M. Catchmark, Je-Ruei Liu, Ali Demirci, and Kuan-Chen Cheng. "Biosynthesis, production and applications of bacterial cellulose." *Cellulose* 20, no. 5 (2013): 2191-2219.

Lin, Wen-Chun, Chun-Chieh Lien, Hsiu-Jen Yeh, Chao-Ming Yu, and Shan-hui Hsu. "Bacterial cellulose and bacterial cellulose–chitosan membranes for wound dressing applications." *Carbohydrate polymers* 94, no. 1 (2013): 603-611.

Liu, Chao, Bin Li, Haishun Du, Dong Lv, Yuedong Zhang, Guang Yu, Xindong Mu, and Hui Peng. "Properties of nanocellulose isolated from corncob residue using sulfuric acid, formic acid, oxidative and mechanical methods." *Carbohydrate polymers* 151 (2016): 716-724.

Lu, Peng, Huining Xiao, Weiwei Zhang, and Glen Gong. "Reactive coating of soybean oil-based polymer on nanofibrillated cellulose film for water vapor barrier packaging." *Carbohydrate polymers* 111 (2014): 524-529.

Lv, XiangGuo, JingXuan Yang, Chao Feng, Zhe Li, ShiYan Chen, MinKai Xie, JianWen Huang, HongBin Li, HuaPing Wang, and YueMin Xu. "Bacterial cellulose-based biomimetic nanofibrous scaffold with muscle cells for hollow organ tissue engineering." *ACS Biomaterials Science & Engineering* 2, no. 1 (2015): 19-29.

Malda, Jos, Jetze Visser, Ferry P. Melchels, Tomasz Jüngst, Wim E. Hennink, Wouter JA Dhert, Jürgen Groll, and Dietmar W. Hutmacher. "25th anniversary article: engineering hydrogels for biofabrication." *Advanced materials* 25, no. 36 (2013): 5011-5028.

Maneerung, Thawatchai, Seiichi Tokura, and Ratana Rujiravanit. "Impregnation of silver nanoparticles into bacterial cellulose for antimicrobial wound dressing." *Carbohydrate polymers* 72, no. 1 (2008): 43-51.

Mao, Angelo S., and David J. Mooney. "Regenerative medicine: current therapies and future directions." *Proceedings of the National Academy of Sciences* 112, no. 47 (2015): 14452-14459.

Markstedt, Kajsa, Athanasios Mantas, Ivan Tournier, Héctor Martínez Ávila, Daniel Hägg, and Paul Gatenholm. "3D bioprinting human chondrocytes with nanocellulose–alginate bioink for cartilage tissue engineering applications." *Biomacromolecules* 16, no. 5 (2015): 1489-1496.

Moon, Robert J., Ashlie Martini, John Nairn, John Simonsen, and Jeff Youngblood. "Cellulose nanomaterials review: structure, properties and nanocomposites." *Chemical Society Reviews* 40, no. 7 (2011): 3941-3994.

Nasir, Mohammed, Rokiah Hashim, Othman Sulaiman, and Mohd Asim. "Nanocellulose: Preparation methods and applications." In *Cellulose-Reinforced Nanofibre Composites*, pp. 261-276. Woodhead Publishing, 2017.

Nimeskern, Luc, Héctor Martínez Ávila, Johan Sundberg, Paul Gatenholm, Ralph Müller, and Kathryn S. Stok. "Mechanical evaluation of bacterial nanocellulose as an implant material for ear cartilage replacement." *Journal of the mechanical behavior of biomedical materials* 22 (2013): 12-21.

Osanlou, Fatemeh, Firouzeh Nemati, and Samaneh Sabaqian. "An eco-friendly and magnetized biopolymer cellulose-based heterogeneous acid catalyst for facile synthesis of functionalized pyrimido [4, 5-b] quinolines and indeno fused pyrido [2, 3-d] pyrimidines in water." *Research on Chemical Intermediates* 43, no. 4 (2017): 2159-2174.

Pavaloiu, Ramona-Daniela, Marta Stroescu, O. A. N. A. Parvulescu, and Tanase Dobre. "Composite hydrogels of bacterial cellulose-carboxymethyl cellulose for drug release." *Rev Chim* 65 (2014): 852-855.

Pereira, Diana R., Joana Silva-Correia, Joaquim M. Oliveira, Rui L. Reis, Abhay Pandit, and Manus J. Biggs. "Nanocellulose reinforced gellan-gum hydrogels as potential biological substitutes for annulus fibrosus tissue regeneration." *Nanomedicine: Nanotechnology, Biology and Medicine* 14, no. 3 (2018): 897-908.

Place, Elsie S., Nicholas D. Evans, and Molly M. Stevens. “Complexity in biomaterials for tissue engineering.” *Nature materials* 8, no. 6 (2009): 457.

Reddy, Jeevan Prasad, and Jong-Whan Rhim. "Isolation and characterization of cellulose nanocrystals from garlic skin." *Materials Letters* 129 (2014): 20-23.

Robson, Anthony. “Tackling obesity: can food processing be a solution rather than a problem?.” *Agro-food-Industry Hi Tech*23, no. 2 (supplement) (2012): 10-11.

Roman, Maren, and William T. Winter. “Effect of sulfate groups from sulfuric acid hydrolysis on the thermal degradation behavior of bacterial cellulose.” *Biomacromolecules* 5, no. 5 (2004): 1671-1677.

Römling, Ute, and Michael Y. Galperin. “Bacterial cellulose biosynthesis: diversity of operons, subunits, products, and functions.” *Trends in microbiology* 23, no. 9 (2015): 545-557.

Ross, Peter, Raphael Mayer, and Moshe Benziman. “Cellulose biosynthesis and function in bacteria.” *Microbiology and Molecular Biology Reviews* 55, no. 1 (1991): 35-58.

Saba, Naheed, and Mohammad Jawaid. “Recent advances in nanocellulose-based polymer nanocomposites.” In *Cellulose-Reinforced Nanofibre Composites*, pp. 89-112. Woodhead Publishing, 2017.

Serizawa, Takeshi, Toshiki Sawada, and Masahisa Wada. “Chirality-specific hydrolysis of amino acid substrates by cellulose nanofibers.” *Chemical Communications* 49, no. 78 (2013): 8827-8829.

Serpa, A., J. Velásquez-Cock, P. Gañán, C. Castro, L. Vélez, and R. Zuluaga. “Vegetable nanocellulose in food science: A review.” *Food Hydrocolloids* 57 (2016): 178-186.

Shaabani, Ahmad, Sajjad Keshipour, Mona Hamidzad, and Mozhdeh Seyyedhamzeh. “Cobalt (II) supported on ethylenediamine-functionalized nanocellulose as an efficient catalyst for room temperature aerobic oxidation of alcohols.” *Journal of chemical sciences* 126, no. 1 (2014): 111-115.

Sheu, F., C. L. Wang, and Y. T. Shyu. "Fermentation of Monascus purpureus on Bacterial Cellulose-nata and the Color Stability of Monascus-nata Complex." *Journal of food science* 65, no. 2 (2000): 342-345.

Singh, Akhilesh V. "Biopolymers in drug delivery: a review." *Pharmacologyonline* 1 (2011): 666-674.

Singla, Rubbel, Sourabh Soni, Pankaj Markand Kulurkar, Avnesh Kumari, S. Mahesh, Vikram Patial, Yogendra S. Padwad, and Sudesh Kumar Yadav. "*In situ* functionalized nanobiocomposites dressings of bamboo cellulose nanocrystals and silver nanoparticles for accelerated wound healing." *Carbohydrate polymers* 155 (2017): 152-162.

Ström, Göran, Camilla Öhgren, and Mikael Ankerfors. "Nanocellulose as an additive in foodstuff." *Innventia Rep* 403 (2013): 1-25.

Sun, J. X., X. F. Sun, H. Zhao, and R. C. Sun. "Isolation and characterization of cellulose from sugarcane bagasse." *Polymer degradation and stability* 84, no. 2 (2004): 331-339.

Syverud, K. "Tissue engineering using plant derived cellulose nanofibrils (CNF) as scaffold material." *Nanocelluloses, their preparation, properties, and applications.* ACS Books, Washington (2017).

Théron, Christophe, Audrey Gallud, Carole Carcel, Magali Gary-Bobo, Marie Maynadier, Marcel Garcia, Jie Lu, Fuyuhiko Tamanoi, Jeffrey I. Zink, and Michel Wong Chi Man. "Hybrid Mesoporous Silica Nanoparticles with pH-Operated and Complementary H-Bonding Caps as an Autonomous Drug-Delivery System." *Chemistry–A European Journal* 20, no. 30 (2014): 9372-9380.

Thomas, Bejoy, Midhun C. Raj, Jithin Joy, Audrey Moores, Glenna L. Drisko, and Clément Sanchez. "Nanocellulose, a versatile green platform: from biosources to materials and their applications." *Chemical reviews* 118, no. 24 (2018): 11575-11625.

Thomas, S., S. A. Paul, L. A. Pothan, and B. Deepa. "Natural fibres: structure, properties and applications." In *Cellulose fibers: bio-and nano-polymer composites*, pp. 3-42. Springer, Berlin, Heidelberg, 2011.

Tummala, Gopi Krishna, Ramiro Rojas, and Albert Mihranyan. "Poly (vinyl alcohol) hydrogels reinforced with nanocellulose for ophthalmic applications: general characteristics and optical properties." *The Journal of Physical Chemistry B* 120, no. 51 (2016): 13094-13101.

Turbak, Albin F., Fred W. Snyder, and Karen R. Sandberg. "Microfibrillated cellulose, a new cellulose product: properties, uses, and commercial potential." In *J. Appl. Polym. Sci.: Appl. Polym. Symp.;(United States)*, vol. 37, no. CONF-8205234-Vol. 2. ITT Rayonier Inc., Shelton, WA, 1983.

Uhrich, Kathryn E., Scott M. Cannizzaro, Robert S. Langer, and Kevin M. Shakesheff. "Polymeric systems for controlled drug release." *Chemical reviews* 99, no. 11 (1999): 3181-3198.

Van den Berg, Otto, Jeffrey R. Capadona, and Christoph Weder. "Preparation of homogeneous dispersions of tunicate cellulose whiskers in organic solvents." *Biomacromolecules* 8, no. 4 (2007): 1353-1357.

Velnar, Tomaž, Tracey Bailey, and Vladimir Smrkolj. "The wound healing process: an overview of the cellular and molecular mechanisms." *Journal of International Medical Research* 37, no. 5 (2009): 1528-1542.

Wang, Hongxia, Jun Qian, and Fuyuan Ding. "Emerging chitosan-based films for food packaging applications." *Journal of agricultural and food chemistry* 66, no. 2 (2018): 395-413.

Wei, Haoran, Katia Rodriguez, Scott Renneckar, and Peter J. Vikesland. "Environmental science and engineering applications of nanocellulose-based nanocomposites." *Environmental Science: Nano* 1, no. 4 (2014): 302-316.

Winter, George D. "Formation of the scab and the rate of epithelization of superficial wounds in the skin of the young domestic pig." *Nature* 193, no. 4812 (1962): 293.

Wu, Xiaodong, Canhui Lu, Wei Zhang, Guiping Yuan, Rui Xiong, and Xinxing Zhang. "A novel reagentless approach for synthesizing cellulose nanocrystal-supported palladium nanoparticles with enhanced catalytic performance." *Journal of Materials Chemistry A* 1, no. 30 (2013): 8645-8652.

Xi Loh, Evelyn Yun, Mh Busra Fauzi, Min Hwei Ng, Pei Yuen Ng, Shiow Fern Ng, Hidayah Ariffin, and Mohd Cairul Iqbal Mohd Amin. "Cellular and Molecular Interaction of Human Dermal Fibroblasts with Bacterial

Nanocellulose Composite Hydrogel for Tissue Regeneration." *ACS applied materials & interfaces* 10, no. 46 (2018): 39532-39543.

Xiang, Qian, Y. Y. Lee, Pär O. Pettersson, and Robert W. Torget. "Heterogeneous aspects of acid hydrolysis of α-cellulose." In *Biotechnology for fuels and chemicals*, pp. 505-514. Humana Press, Totowa, NJ, 2003.

Yu, Houyong, Zongyi Qin, Banglei Liang, Na Liu, Zhe Zhou, and Long Chen. "Facile extraction of thermally stable cellulose nanocrystals with a high yield of 93% through hydrochloric acid hydrolysis under hydrothermal conditions." *Journal of Materials Chemistry A* 1, no. 12 (2013): 3938-3944.

Zhao, Jiangqi, Canhui Lu, Xu He, Xiaofang Zhang, Wei Zhang, and Ximu Zhang. "Polyethylenimine-grafted cellulose nanofibril aerogels as versatile vehicles for drug delivery." *ACS applied materials & interfaces* 7, no. 4 (2015): 2607-2615.

Biographical Sketch

Avnesh Kumari

Affiliation: CSIR-IHBT

Education: MSc, PhD

Business Address: Nanobiology Lab, CSIR-IHBT, Palampur (H.P.)-India-176061

Research and Professional Experience: 12 years experience in the area of nanobiology and electron microscopy. Presently working towards improving the solubility, bioavailability and efficacy of important phytomolecules possessing activities like antioxidant, antimicrobial and anticancer. Secondly also working on the isolation, and characterization of nanocellulose from bio resources.

Professional Appointments:

Scientist Fellow: 16 June 2008 to 31st August, 2009
Principal Investigator: 1st September, 2009 to 12th January, 2011
Senior Technical Officer (2): 13th January, 2011 to Till date

Honors: Most cited and most downloaded article of the Journal
Biodegradable polymeric nanoparticles based drug delivery systems.
Kumari A, Yadav SK, Yadav SC. Colloids Surf B Biointerfaces. 2010 Jan 1;75(1):1-18. doi: 10.1016/j.colsurfb.2009.09.001. Epub 2009 Sep 8. Review. (IF = 4.2)

2162 citations
Publications from the Last 3 Years:

Research articles

1. Anika Guliani, Rubbel Singla, Avnesh Kumari and Amitabha Acharya (2018) Effect of surfactants on the improved selectivity and anti-bacterial efficacy of citronellal nano-emulsion, *Journal of Food Process Engineering*, (In press).
2. Singla R, Soni S, Patial V, Kulurkar P, Kumari A, Mahesh S., Padwad Y, and Yadav S, Cytocompatible Anti-mircobial Dressings of Syzygium cumini Cellulose Nanocrystals Decorated with Silver Nanoparticles Accelerate Acute and Diabetic Wound Healing, *Nature Scientific reports*, 2017, 7,10457 (IF = 4.5)
3. Singla S, Soni S, Patial V, Kulurkar P, Kumari A, Mahesh S.,Padwad YS,Yadav SK, *In vivo* diabetic wound healing potential of nanobiocomposites containing bamboo cellulose nanocrystals impregnated with silver nanoparticles, *International Journal of Biological Macromolecules*,2017, 105,45-55 (IF =4.78).
4. Singla, R., Soni, S., Kulurkar, P. M., Kumari, A., Mahesh, S., Patial, V., Padwad YS & Yadav, S. K. (2017). *In situ* functionalized nanobiocomposites dressings of bamboo cellulose nanocrystals and

silver nanoparticles for accelerated wound healing. *Carbohydrate Polymers*, *155*, 152-162. (**IF = 6.1**)

Book chapters

1. Kumari A, Singla R, Anika Guliani A and Yadav SK, Biodegradable nanoparticles and their *in vivo* fate, in *Nanoscale materials in targeted drug delivery, theragnosis and tissue regeneration,* Springer, 2016.
2. Kumari A, Singla R, Anika Guliani A and Yadav SK, Cellular response of therapeutic nanoparticles in *Nanoscale materials in targeted drug delivery, theragnosis and tissue regeneration*, Springer, 2016.
3. Kumari A, Singla R, Anika Guliani A and Yadav SK, Nanoscale materials in targeted drug delivery in *Nanoscale materials in targeted drug delivery, theragnosis and tissue regeneration,* Springer, 2016.
4. Guliani A, Singla R, Kumari A and Yadav SK, Liposomal and phytosomal formulations, in *Nanoscale materials in targeted drug delivery, theragnosis and tissue regeneration*, Springer, 2016.
5. Singla R, Guliani A, Kumari A and Yadav SK, Nanocellulose and nanocomposites, in *Nanoscale materials in targeted drug delivery, theragnosis and tissue regeneration*, Springer, 2016.
6. Singla R, Anika Guliani A, Kumari A and Yadav SK, Metallic nanoparticles, toxicity issues and applications in medicine, in *Nanoscale materials in targeted drug delivery, theragnosis and tissue regeneration*, Springer, 2016.
7. Singla R, Guliani A, Kumari A, Yadav SK Role of Bacteria in Nanocompounds formation and their Application in Medical, in *Microbial Applications* Vol.2 pp 3-37, Springer, 2016.
8. Kumari, A, Kumar V and Yadav, SK, The use of *Syzygium Cumini* in nanotechnology, in *The Genus Syzygium Cumini and other underutilized specie Nair KN* (Eds.) 2017,pp-171-195, CRC press.

Patents

1. A nanobiocomposite formulation for wound healing and a process for the preparation thereof, Ref No- 0916NF2016, Filed in US, Japan and Europe.
   Inventors: Sudesh Kumar Yadav, Avnesh Kumari and Rubbel Singla
2. Plant extract based process for the synthesis of PLA nanoparticles and their use for controlled and sustained release of poor water soluble molecules, 0040NF2017.
   Inventors: Sudesh Kumar Yadav, Avnesh Kumari and Anika Guliani
3. A green process for the synthesis of curcumin loaded PLGA nanoparticles with increased solubility and photo - stability, 0066NF2018.
   Inventors: Amitabha Acharya, Avnesh Kumari, Anika Guliani and Sanjay Kumar

In: Cellulose Nanocrystals
Editor: Orlene Croteau
ISBN: 978-1-53616-747-4

Chapter 4

CELLULOSE NANOCRYSTALS AND ITS APPLICATION IN POLYMER NANOCOMPOSITE FOR DRUG DELIVERY

Jonathan Tersur Orasugh[1,2,3], Swapan Kumar Ghosh[2,3*] and Dipankar Chattopadhyay[1,3,*]

[1]Department of Polymer Science and Technology,
University of Calcutta, West Bengal, India
[2]Department of Jute and Fibre Technology,
Institute of Jute Technology, University of Calcutta, West Bengal, India
[3]Center for Research in Nanoscience and Nanotechnology,
Acharya Prafulla Chandra Roy Sikhsha Prangan,
University of Calcutta, Saltlake City, Kolkata

ABSTRACT

Cellulose is the most abundant biopolymer on earth has been applied in almost all areas/field of material science and engineering with cellulose nanocrystals (CNC) being one of the highly researched/applied form of

[*]Corresponding Author's Email:dipankar.chattopadhyay@gmail.com.

cellulose recently. This book chapter gives a brief up to date introductory review on CNC, its synthesis approach, up to date research in polymer-based CNC nanocomposites, its applications, challenges and scope for future applications and conclusion. The chapter serves as a guide for researchers, scientist, engineers, and technologist to know areas where there is still room for utilization of CNC in advance materials applications and also to effectively develop/proffer new novel application areas for CNC.

Keywords: cellulose, CNC, nanocomposite, material science, polymer, drug delivery

1. Introduction to Cellulose and CNC in Drug Delivery

Within the last few decades, noteworthy progress has been made in the field of drug delivery with the advancement in sustained release of drugs. There is a huge variety of pharmaceutical formulations committed to oral sustained delivery of the drug. Drug release modes can be categorized into those that deliver the drugs at an unhurried zero or first-order rate and those that provide an initial rapid release, followed by sluggish zero or first-order release of drug entities [1]. Sustained drug delivery systems (SDDS) are aimed at maintaining drug steady concentration of the administered drugs at the target site or in the blood plasma resulting in increased bioavailability [1, 2]. SDDSs are the modern and novel model in therapeutic treatment aimed at increase drug efficiency and patient conformity by reducing the drug administration frequency and side effects [1, 3].

Bio-renewable materials synthesized from natural sources such as CNC have gained widespread exploration because of their many potential applications as biomaterials in drug delivery [4-11]. This has led to the drive by researchers worldwide towards facile development of site-specific nontoxic novel drug carriers with high therapeutic bioavailability [4-16]. Within the last decade, the design of the aforementioned novel carriers for advanced technical applications using renewable resources has directed the researcher's search beam towards renewable biopolymers which are nontoxic, bio-absorbable, biodegradable, biocompatible, low density, high strength properties, and possess the potential for chemical modification.

It is generally known that cellulose fiber is the most abundant renewable biomaterial in the biosphere [4, 17-26]. The hydrolysis of cellulose fibres using acids (H_2SO_4 or HNO_3) disrupts the amorphous regions of cellulose leading/resulting to the releases of individual rod/rice-like rigid crystallites referred to as cellulose nanocrystal (CNC) or cellulose nanowhiskers which possess excellent mechanical properties [4,17-26].CNC which was first discovered and reported in 1949 [27], has been of high importance within recent decades due to its availability, high surface area and aspect ratio along with non-cytotoxicity, biocompatibility and biodegradability character. These properties present CNC as a potential material of choice in biomaterial science and technology [4, 5, 17, 19-23, 25-27]. The high surface area coupled with a high number of hydrophilic (hydroxyl) groups makes it possible for a huge quantity of drugs attached/loaded onto CNC and also better-sustained release of the drugs in both ophthalmic and transdermal drug delivery systems [7-11]. The rod-like CNCs and rice/needle-like CNCs are both prepared using controlled acid hydrolysis of pure cellulose. In general, CNC have been sourced from jute fibres [7-11], bacterial cellulose [28], Wood [29, 30, 31], sugar beet primary wall cellulose [32, 33], cotton [8, 34, 35], tunicate [36, 37, 38], along with wheat straw [39-41]. Factors such as the cellulose source, depolymerization/hydrolysis parameters, and ionic strength of the solution/solvents influence the size of the isolated CNC [42]. The size of CNC as reported from previous literature includes 7.1–7.9 nm x 95–114 nm & 7 × 100–300 for cotton [8,43],10–15 nm x ~89 nm, & ~15–28 nm x ~90–134 nm for jute fibres [8-11], 3–5 × 180 ± 75 nm for wood [39], 5–10 nm × 100 nm to several μm for BC [28], as well as 10–20 nm × 100 nm to several μm for tunicate [44].

The application of CNC based nanocomposites in drug delivery application has been deeply studied by very few researchers around the globe [45]. Akhlaghi and co reported oxidized CNC (CNC-OX) and chitosan oligosaccharide grafted CNC and evaluated the same as potential drug delivery carriers for two model drug compounds procaine hydrochloride (PrHy) and imipramine hydrochloride (IMI)) at pH 8 and 7 respectively where IMI displayed higher binding to CNC derivatives than PrHy due to the more dominant heat change observed in the ITC isotherms of the drug-loaded nanocomposites of IMI. The release kinetics of CNC polymer matrix based systems reveals dependence on the solubility of the drug in the medium and its affinity to CNC [45]. CNC based biopolymeric matrix nanocomposites have been engineered for the sustained release of metronidazole [16], alfacalcidol [46], procaine hydrochloride [45], theophylline [47], ibuprofen [48, 49], propranolol hydrochloride [50], curcumin [51], procaine hydrochloride [45], doxorubicin [52], metronidazole [16], curcumin [53], pilocarpine hydrochloride [11], ketorolac and tromethamine [54]. All the reports demonstrated the CNC has a good impact on the drug release kinetics of the systems. Biopolymers and their blends such as CNC-CTS [45], CNC [48], poly(N-isopropylacrylamide) (PNIPAAm) [16], alginate [46], poly(acrylic acid) (PAA) [47], alginate, polyacrylic acid [49], collagen-CNC [50], CTS-CNC [51], pluronic [11,45], poly(N-isopropylacrylamide) (PNIPAAm) [52], chitosan (CS) [16] polyvinylpyrrolidone (PVP) [52], and methylcellulose (MC) [10] have been engaged for the design and engineering of CNC based drug delivery devices for enhanced bioavailability.

In recent years, more than few reviews have summarized the synthesis and exploitation of nanocellulose based materials in general for applications in edible packaging, biomaterials, filtration, tissue engineering, adsorption, paper, electronics, etc which channels the readers towards the aforementioned areas [55, 56, 57]. However, highly structured summaries essentially aimed at the application of CNC based nanocomposites in drug delivery are either rare or limited. As a result, the present book chapter, for the most part, focuses on the recent emerging areas of application of drug-loaded CNC-biopolymer nanocomposites in drug delivery systems.

2. PROPERTIES OF CNC

The properties of CNC as presented in literatures have shown that it possess excellent properties such as high tensile strength (TS) [44], high crystallinity index [8-11], excellent axial stiffness, low coefficient of thermal expansion (CTE), excellent thermal stability (THS) [58], high barrier properties [8-11,58], high aspect ratio, low density [8-11,58], shear-thinning & lyotropic rheological properties, good zeta potential for effective dispersion [8-11], high surface area (densities = 0.01–0.4 g/cm3, surface area = 30–600 m^2/g) [58,59], and higher number of surface -OH groups for easy modification as presented in Table 1. The attractive eco-friendly characteristics of CNCs include; negligible environmental impact, acceptable health conformity, no safety risks, renewable, sustainable, biocompatible, bioabsorbable, and easy to be isolated at low industrial-scale. CNCs are characterized by a degree of polymerization (DP) ranging from $90 \leq DP \leq 110$, and having 3.7-6.7 sulfate groups adsorbed per 100 anhydroglucose units along with a high degree of crystallinity (greater than 80%) for H_2SO_4 isolated CNCs [59].

Table 1. Selected properties of CNC

Tensile Strength (GPa)	Stiffness (GPa)	Dimension (nm)	Thermal		Density (g/cm^3)	Rheological	Zeta potential (mV)	Crystallinity (%)	Aspect ratio
			CTE (ppm/K)	THS (°C)					
≥7.5 - 7.7	≥150	1, 3 – 20 x 50–2000	~1	~300	~1.6	Lyotropic & shearthinning	-12 to - 12.6	≥80	7 ≤100

3. Cellulose Nanocrystals and Its Nanocomposites in Drug Delivery

The high surface area enhanced surface –OH groups on the CNCs provide a readily active surface for interaction with other engineered hydrophilic matrix surface groups in the formation of highly compatible nanocomposites via excellent interfacial adhesion via non covalent bonding along with potential modification such as cross-linking or grafting between CNCs in the matrix or CNCs and the matrix. The transparency, thermal stability, nontoxic, low density, excellent mechanical properties, and nonpermeable nature of CNCs make them very much attractive in preparing nanocomposite materials. These distinctive qualities of CNCs open doors to new opportunities in comparison with conventional lignocellulose/cellulose fillers like jute fibres, ramie fibres, wood flakes, pulp fibres, etc [5, 6, 17-19, 20-24, 26, 27, 58].

Table 2. Report of drug-loaded CNC-bionanocomposites

Matrix used	CNC source	Enhanced properties	Loaded drug	Release mechanism	Proposed Application	References
BIS/PAM	wood	Mechanical				61
CNC-CTS	Supplied by FP Innovations		procaine hydrochloride		Wound dressing & TDDS	59
CNC	BC	Sustained release	ibuprofen		DD	48
poly(N-isopropylacrylamide) (PNIPAAm)	Whatman–CF11	Rheological properties	metronidazole	Case II kinetics	Wound dressing	16
Alginate	BC	Sustained release	Alfacalcidol	non-Fickian transport	Scaffold & drug delivery	46
poly(acrylic acid) (PAA)	Kenaf fibers	Mechanical, rheological	theophylline	non-Fickian diffusion	Drug delivery	47
alginate	Rice Husk	Sustained release	Ibuprofen	diffusion control	Drug delivery	49
polyacrylic acid	MCC	Sustained release	Propranolol hydrochloride	diffusion control	DD	50
Collagen-CNC	-	Sustained release, Wound healing	Curcumin	-	Wound dressing	51
CTS-CNC	supplied by FP Innovations	Sustained release, Wound healing	procaine hydrochloride	-	Wound dressing (WD)	45
Pluronic	Cotton fibres	Sustained release, rheological	doxorubicin	anomalous transport	DD	52
chitosan (CS) and polyvinylpyrrolidone (PVP)	cellulose powder	Sustained release	curcumin	diffusion-controlled	WD	53
sodium alginate	banana fibre				DD	62

Polymer nanocomposite can be defined as a multiphase material comprising of a very low volume fractions nanomaterial reinforced polymer matrix with unique properties compared to traditional composite materials due to the influence of the reinforcing nanofiller. Biopolymers (natural and synthetic) have been reportedly used for the engineering of novel CNC based nanocomposites for advanced technical applications especially in biopharmaceutical materials as also shown in Table 2 [4-11, 17-19, 20-27, 58, 60].

3.1. Release Mechanism in CNC Based Drug Delivery Systems

The drug release mechanism/kinetics of drug load polymer systems (especially hydrophilic polymers) is greatly influenced by the swelling behavior of the polymer matrix [11]. Rifampicin-loaded alginate-cellulose nanocrystals hybrid nanoparticles have proven to have a superior result for the treatment of *Mycobacterium tuberculosis* in comparison to the pure alginate loaded rifampicin (RIF) where they revealed that RIF release was extremely low for the first 2 h, equivalent to pH 1.2 and the release increased over the subsequently 4 h, at pH 6.8 and increased further before reaching a maximum at pH 7.4 within the next 6 h but they did not mention the release mechanism/kinetics [62]. The adaption of CNCs for sustained topical delivery (up to 80% within 4 h) of hydroquinone aimed at inhibition of the production of melanin and eliminate the discolorations of skin (hyperpigmentation) has been reported where Hydroquinone molecules were probably attached to CNC via a hydrogen bond with the carboxyl groups on the surface of CNC [54]. Solid nanoparticles such as CNC are said to kinetically stabilize Pickering emulsion and foams consist of drops or bubbles kinetically stabilized due to the small solid particles attached to the drop (or bubble) surface [63]. Aimed at understanding the drug release mechanisms of CNC based nanocomposite systems, researchers have adopted Zero order (ZO), First-order (FO), Higuchi's, and Korsmeyer-Peppas (KP) models to analyze the release kinetics of loaded drugs in these systems [7-11, 13, 16, 25, 50, 62, 64, 65]. CNC-MC drug-loaded nanocomposite films have been reported to follow FO kinetics with the linearity of $r^2 > 0.98$ [10]. Yet in another study, the *in vitro* drug release of CNC reinforced PM nanocomposite hydrogels formulations was also explained by the first-order equation, as the plots depicted the highest linearity of $r^2 \geq 0.981$ [11]. Cellulose microcrystal's (CMCs) reinforced chitosan microparticles on the other hand have been reported to follow Higuchi kinetic model [66]. This model describes the absorption and/or elimination of drugs loaded to/onto. However, it is difficult to conceptualize this mechanism on a theoretical basis. The release of the drugs which follow FO kinetics can be articulated as:

$$dC/dt = -K$$

Where K is first-order rate constant expressed in units of time^{-1}. The equation above can also be written in the form:

$$logC = (logC_o - Kt)/2.303$$

Where K, C_0, & C, and t the first-order rate constant, the initial & final concentration of the drug, and the time [67].

On the other hand, Higuchi model which focuses on giving an explanation to matrix-based drug system (Initially conceived for planar systems and then extended to different geometrics and porous systems) as shown below [68]:

$$f_t = Q = A\sqrt{D}(2C - C_s)C_s t$$

Where Q is the amount of drug released in time t per unit area A, C is the drug initial concentration, Cs is the drug solubility in the matrix media and D is the diffusivity of the drug molecules (also known as diffusion coefficient) in the matrix substance. This relation is valid all the time, except when the total depletion of the drug in the therapeutic system is achieved. Where the drug concentration in the matrix is lower than its solubility and the release occurs through pores in the matrix such as the dissolution of the drugs from a planar heterogeneous matrix system, the equation above can be expressed as [69]:

$$f_t = Q = \surd(D\delta/\tau)\ (2\mathrm{C}-(\delta C_s)C_s t$$

Where D is the diffusion coefficient of the drug molecule in the solvent, δ is the porosity of the matrix, τ is the tortuosity of the matrix which is defined as the dimensions of radius and subdivisions of the pores and channels in the matrix. The Higuchi model is generally simplified as [69]:

$$f_t = Q = K_H \times t^{1/2}$$

where K_H is the Higuchi dissolution constant. The above expressing is generally used to describe the drug dissolution/release from different forms of engineered drug release pharmaceutical formulations like TDDS, ODDS and matrix tablets/excipient with hydrophilic water soluble drugs [69, 70]:

4. Application of CNC in Polymer Nanocomposite for Drug Delivery

Nanoscale carrier systems are highly introspected with regards to curative and diagnostic medicine. Thus cellulose nanocrystals may be considered as an ideal nanoscale carrier candidate for bioactive molecules due to its lack of toxicity an untargeted uptake. A large number of drugs can be bound to the surface of CNC due to a negatively charged surface along with its very large surface area. Thus it can be easily converted to other functional groups for covalent and non-covalent binding of molecules with impactful biological effects. So compared to hydrophobic nanoparticles CNC-bound drugs are expected to have an inherently prolonged blood circulation half-life that is the time it takes for removing the half of the drug when the rate of removal is roughly exponential.

The applications of bioderived CNCs includes reinforcing filler, drug carrier, barrier films, food additives, tissue scaffolding, etc. CNCs are also applied as rheology modifiers [11, 13,25]for solutions like *in situ* polymer gels, polymer melts, particle mixtures for utilization in ophthalmic formulations, paints, transdermal patches, coatings, food, injectable gels, adhesives, lacquers, cosmetics, drug excipient, etc. CNCs are also reportedly applied as reinforcing filler in polymer matrix aimed at altering the mechanical behavior of the consequential nanocomposites for the preparation of tough, flexible, biodegradable, durable, lightweight, transparent, and dimensionally stable nanomaterial.

CNC-based polymer film provides well-mannered barrier character due to the impermeable nature of the crystals to the diffusion of gases, liquids and to UV radiation into the drug-loaded CNC-based nanocomposite material [9-10]. The creation of a tortuous path in the polymer matrix based nanocomposite also enhances the barrier properties in these materials [9-11, 71]. Figure 1 demonstrates the barrier features of the CNC based polymeric nanocomposite (particles 1-17).

Application of CNC in TDDS is ideally suitable in case of diseases which need chronic treatment. This is gaining its importance as an alternative to oral medicines and hypodermic injections [72-74]. The basic principle of this technique is the development of a transdermal patch that is placed on the skin in order to deliver a particular dose of medication through the skin and into the bloodstream [72-74]. Now, with microneedle gaining enhancement in transdermal drug delivery [73], cellulose nanocrystal being known for its non-toxicity may be used to fabricate the carries in the microneedles TDDS systems. These small needles penetrate the top layer of the skin and make way for the drugs to pass with ease thereby placing the drugs in a precise manner into the region where the immune cells are residing for easy to modulation of the immune system [73].

The controlled delivery of theophylline from a novel fabricated CNC/CS composites presenting cumulative drug release percentage of the nanocomposite hydrogel of ~85% and ~23% in the gastric (pH 1.5) and intestinal (pH 7.4) fluids have reported [75] which presents the application of CNC in ODDS as an expansion of new method for ocular drug delivery which can lead to the treatment of challenging diseases in the anterior and posterior segment of the eye [75]. One of the foremost areas of attraction for the utilization of CNC at present is the biomedical and biopharmaceutics considering its biocompatibility, ease of chemical modifications, nontoxic nature, inexpensive source, costs of production, exceptional tensile properties, sustainability, bioabsorbability, and high surface area [9-10, 13, 25].

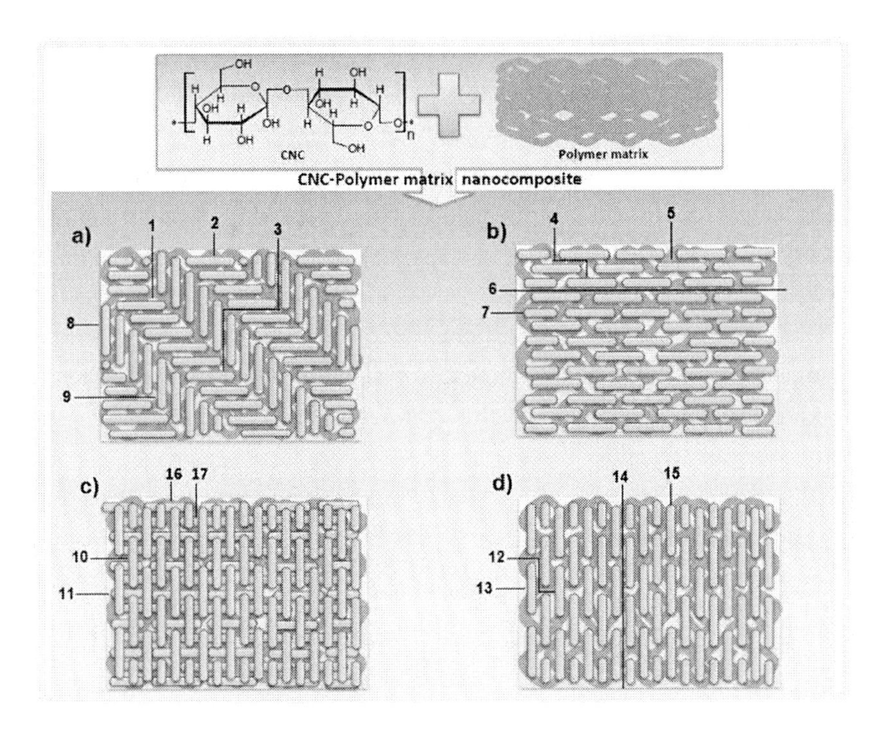

Figure 1. Schematic demonstration the barrier features of the CNC based polymeric nanocomposite.

The hydrogel-forming ability arising from high water holding capacity coupled with the morphology of CNC which presents a highly porous three-dimensional structure suitable for the culturing of cells in tissue scaffold engineering for enhanced cell proliferation has been presented [9-10, 13, 25].

The low concentration hydrogel-forming character of CNC has been utilized for the reduction of the critical gelation concentration of smart polymer-based hydrogels having the potential for application in novel ophthalmic formulations with enhancing drug delivery [11]. CNC has been shown to interact with the drugs resulting in the stability of the drug particles and subsequently sustained release of the drugs.

Another novel area of application where CNC has proven feasible is edible packaging which is very vital in the pharmaceutical industry [9-10, 13, 25]. CNC enhance the properties of these materials with negligible effect on the transparency and weight of the final composite due to their nano-sized particle and low density [20-24].

The controlled digestion of protein-based nanocomposite for *in vitro* gastric digestion of emulsions using composite whey protein-cellulose nanocrystals interfaces resulting in delayed protein-particle interface gastric digestion have been reported [63].

CNC as bioderived nanoparticles/filler acts as a reinforcing filler in polymer matrix based nanocomposite system thereby strengthening the material even at very low volume fraction and also replacing non environmentally friendly fibres or synthetic origin has been proven effective [9-10]. Other applications of CNC are filtration membranes, oil recovery, dye adsorption, etc.

5. Present Challenges and Future Scope

The future utilization of CNC nanocomposite based nanomaterials in biomaterials engineering looks very promising. The isolation approach of CNC may be improved upon with focus on green extraction methods while some of the facile novel techniques can be preserved [76]. To attract the advanced application of CNC in DD, nontoxic modification routes with the proven industrial application should be adopted [77, 78]. There is room for the application of engineered CNCs for as potential starting material for designing biomarkers for early detection of terminal/non-terminal diseases. Like liposome's, CNCs may possess a probability of damaging the vitreous (gel that fills the space between the lens and retina of the eyeball) and also since its long terms effects are not understood, they are yet to be established as an alternative medication in ODDS. The market prospect for CNC in the turf of biomaterials, pharmaceuticals, water treatment, energy, agriculture, and electronics are massive, nevertheless, there are innovative opportunities in more worthy areas. Due to the benevolent uniqueness of CNCs coupled with their toxic nature, much prospect exists for the exploit of CNCs in cosmetics, bioimaging, and engineered edible food materials to reduce the impact of toxic materials on the ecosystem. The next decades will see a surplus of innovations in CNC based sustainable nanomaterials considering the current drive of researchers worldwide toward the integration of CNC based materials aimed at replacing non-renewable materials with renewable/sustainable materials. Indisputably, CNC possesses enormous properties for advance engineering of a novel new cohort of nano-engineered materials for application in almost all branches of material science and technology. Research studies on CNC have shown development with the guarantee of even superior progress probable in-store within decades to come. There is a need for more comparative research on the properties of BC, CNF, and CNC, to ascertain their influence on the overall properties of the nanocomposite in order to strategize their effective utilization in bionanocomposite for drug delivery applications. Moreover, the study of the interaction between CNCs to loaded drugs, CNCs to cultured cells, CNCs to the polymer matrix, and CNC-CNC need further analysis. The research on the progress in CNC based drug delivery systems with enhanced/better bioavailability compared to target at promoting the beginning of numerous money-making products with proven real-life usage in biomaterials. As Mother Nature's gift, CNC as a novel form of cellulose is on the verge of an exceptional attraction in modern astonishing progress in the niche of biomaterial applications.

ACKNOWLEDGMENTS

The authors appreciate the Centre for Research in Nanoscience and Nanotechnology (CRNN), Department of Polymer Science and Technology, and Department of Jute and Fibre Technology, Institute of Jute Technology, the University of Calcutta for supporting this work.

REFERENCES

[1] Langer R.S., Wise D.L. 1984. *Medical applications of controlled release, applications and evaluation.* Eds. Vol. I and II, CRC Press, Boca Raton.

[2] Abo-Elseoud S.W., Hassan L.M., Saba W.M., Basha M., Hassan A.E., Fadel M.S. 2018. Chitosan nanoparticles/cellulose nanocrystals nanocomposites as a carrier system for the controlled release of repaglinide: *International Journal of Biological Macromolecules,* 111:604-613.

[3] Robinson J.R., Lee, V.H.L. 1987. *Controlled drug delivery*. Eds. Marcel Dekker, Inc. New York, Basel.

[4] Yang J., Han R. C., Duan F. J., Ma G.M., Zhang M. X., Xu F., Sun C. R. 2013. Synthesis and characterization of mechanically flexible and tough cellulose nanocrystals–polyacrylamide nanocomposite hydrogels. *Cellulose,* 20:227–237.

[5] Yang J., Han R. C, Duan F. J., Xu F., Sun C. R. 2013. Mechanical and Viscoelastic Properties of Cellulose Nanocrystals Reinforced Poly (ethylene glycol) Nanocomposite Hydrogels. *ACS Applied Materials & Interfaces*, 5:3199−3207.

[6] De Moura M.R., Avena-Bustillos R.J., Mchugh T.H., Wood D.F., Otoni C.G., Matteson. H.C. 2011. Miniaturization of cellulose fibers and effect of addition on the mechanical and barrier properties of hydroxypropyl methylcellulose films. *Journal of Food Engineering* 104: 154–166.

[7] Dutta D., Das B., Orasugh T. J., Mondal D., Adhikari A., Rana D., Banerjee R, Mishra R, Kar, S, Chattopadhyay D. 2018. Bio-derived cellulose nanofibril reinforced poly (N-isopropylacrylamide)-g-guar gum nanocomposite: An avant-garde biomaterial as a transdermal membrane. *Polymer*, 135: 85-102.

[8] Orasugh J.T., Saha N.R., Sarkar G., Rana D., Mondal D., Ghosh K.S., Chattopadhyay D. 2018a. A facile comparative approach towards utilization of waste cotton lint for the synthesis of nano-crystalline cellulose crystals along with acid recovery. *International Journal of Biological Macromolecule,*109:1246-1252.

[9] Orasugh J.T., Saha R.N., Rana D., Sarkar G., Mollick R.M.M, Chattopadhyay A., Mitra C.B., Mondal D, Ghosh K.S., Chattopadhyay D. 2018b. Jute cellulose nano-fibrils/hydroxypropylmethylcellulose nanocomposite: A novel material with potential for application in packaging and transdermal drug delivery system. *Industrial Crops and Products, 112:633*-643.

[10] Orasugh J.T., Saha R.N., Sarkar G., Rana D., Mishra R., Mondal D., Ghosh K.S., Chattopadhyay, D. 2018c. Synthesis of methylcellulose/ cellulose nano-crystals nanocomposites: Material properties and study of sustained release of Ketorolac tromethamine. *Carbohydrate Polymers,* 188:168-180.

[11] Orasugh J.T., Sarkar G., Saha N.R., Das B., Bhattacharyya A., Das S., Mishra R., Roy I., Chattopadhyay A., Ghosh S.K., Chattopadhyay D. 2019. Effect of cellulose nanocrystals on the performance of drug loaded *in situ* gelling thermo-responsive ophthalmic formulations. *International Journal of Biological Macromolecule,* 124:235–245.

[12] Oh Y.S., Yoo I. D., Shin Y., Kim C.H., Kim Y.H., Chung S.Y., Park H.W., and Youk H.J. 2005. Crystalline structure analysis of cellulose treated with sodium hydroxide and carbon dioxide by means of X-ray diffraction and FTIR spectroscopy. *Carbohydrate Research,* 340:2376–2391.

[13] Yu Y. H., Wang C., Abdalkarim H.Y.S. 2017. Cellulose nanocrystals/polyethylene glycol as bifunctional reinforcing/

compatibilizing agents in poly (lactic acid) nanofibers for controlling long-term in vitro drug release. *Cellulose*, 24:4461–4477.

[14] Bhowmik M., Kumari P., Sarkar G., Bain K.M., Bhowmick B., Mollick R.M.M, et al. 2013. Effect of xanthan gum and guar gum on in situ gelling ophthalmic drug delivery system based on poloxamer-407. *International Journal of Biological Macromolecule,* 62: 117-123.

[15] Dewan M., Sarkar G., Bhowmick M., Das B., Chattopadhyay K. A., Rana D., et al. 2017. Effect of gellan gum on the thermogelation property and drug release profile of Poloxamer 407 based ophthalmic formulation. *International Journal of Biological Macromolecule,* 102: 258-265.

[16] Zubik K., Singhsa P., Wang Y., Manuspiya H., and Narain R. 2017. Thermo-Responsive Poly (*N*-Isopropylacrylamide)-Cellulose Nanocrystals Hybrid Hydrogels for Wound Dressing. *Polymers. 9*:119.

[17] Zoppe O.J., Habibi Y., Rojas J.O., Venditti A.R., Johansson S-L, Efimenko K., Sterberg O. M., Laine J. 2010. Poly (*N*-isopropylacrylamide) Brushes Grafted from Cellulose Nanocrystals via Surface-Initiated Single-Electron Transfer Living Radical Polymerization. *Biomacromolecules.*11:2683–2691.

[18] Zhang H., Jung J., Zhao Y. 2017. Preparation and characterization of cellulose nanocrystals films incorporated with essential oil loaded β-chitosan beads. *Food Hydrocolloids*.69:164-172.

[19] Yanamala N., Farcas T.M, Hatfield K.M., Kisin R.E., Valerian E. Kagan, Charles L. Geraci, and Anna A. Shvedova. In Vivo Evaluation of the Pulmonary Toxicity of Cellulose Nanocrystals: A Renewable and Sustainable Nanomaterial of the Future. *ACS Sustainable Chemical Engineering*, 2014, 2, 1691−1698.

[20] Wang T. and Drzal T.L. 2012. Cellulose-Nanofiber-Reinforced Poly (lactic acid) Composites Prepared by a Water-Based Approach. *ACS Applied Materials & Interfaces*, 4:5079−5085.

[21] Wang S., Sun J., Jia Y., Yang L., Wang N., Xianyu Y., Chen W., Li X., Cha R., and Xingyu Jiang. 2016. Nanocrystalline Cellulose-Assisted Generation of Silver Nanoparticles for Nonenzymatic

Glucose Detection and Antibacterial Agent. *Biomacromolecules*, 17, 2472−2478.

[22] Wang H., He J., Zhang M., Tam C.K., and Ni P. 2015. A new pathway towards polymer modified cellulose nanocrystals via a "grafting onto" process for drug delivery. *Polymer Chemistry*, 6:4206-4209.

[23] Wang S., Lu A., Zhang L. 2016 Recent advances in regenerated cellulose materials. *Progress in Polymer Science*, 53:169–206.

[24] Zhang P.P., Tong S.D., Lin C.X., Yang M.H., Zhong K.Z., Yu H.W., Wang H., and Zhou H.C. 2014. Effects of acid treatments on bamboo cellulose nanocrystals. *Asia-Pacific Journal of Chemical Engineering*, 9:686–695.

[25] Yu S., Zhang X., Tan G., Tian L., Liu D., Liu Y., Yang X., Pana W. 2017. A novel pH-induced thermosensitive hydrogel composed of carboxymethyl chitosan and poloxamer cross-linked by glutaraldehyde for ophthalmic drug delivery. *Carbohydrate Polymers,* 155: 208-217.

[26] Xiang C., Taylor G.A., Hinestroza P.J., Frey W.M. 2013. Controlled Release of Nonionic Compounds from Poly (lactic acid)/Cellulose Nanocrystal Nanocomposite Fibers. *Journal of Applied Polymer Science*, DOI: 10.1002/APP.36943.

[27] Moreau C., Villares A., Capron I., & Cathala B. 2016. Tuning supramolecular interactions of cellulose nanocrystals to design innovative functional materials. *Industrial Crops and Products*, 93:96–107.

[28] Roman M. and Winter T.W. 2004. Effect of Sulfate Groups from Sulfuric Acid Hydrolysis on the Thermal Degradation Behavior of Bacterial Cellulose. *Biomacromolecules*, 5:671−1677.

[29] Araki J., Wada M., Kuga S., and Okano T. 1999. Influence of Surface Charge on Viscosity Behavior of Cellulose Microcrystal Suspension. *Journal of Wood Science*, 45, 258–261 (1999).

[30] Revol J.F., Bradford H., Giasson J., Marchessault R.H., and Gray D.G. 1992 Helicoidal Self-Ordering of Cellulose Microfibrils in Aqueous Suspension. *International Journal of Biological Macromolecules*, 14:170–172.

[31] Abitbol T., Johnstone T., Quinn T.M., and Gray D.G. 2011. Reinforcement with cellulose nanocrystals of poly (vinyl alcohol) hydrogels prepared by cyclic freezing and thawing. *Soft Matter*, 7(6):2373.

[32] Dufresne A., Cavaille J.Y., and Vignon M.R. 1997. Mechanical Behavior of Sheets Prepared From Sugar Beet Cellulose Microfibrils. *Journal Applied Polymer Science*, 64:1185–1194.

[33] Dinand E., Chanzy H., and Vignon M.R. 1999 Suspensions of Cellulose Microfibrils from Sugar Beet Pulp. *Food Hydrocolloids*, 13:275–283.

[34] Maiti S., Jayaramudu J., Das K., Reddy S.M., Sadiku R., Ray S.S., Liu D. 2013. Preparation and characterization of nano-cellulose with new shape from different precursor. *Carbohydrate Polymers*, 98 562–567.

[35] Spagnol C., Rodrigues F.H.A., Pereira A.G.B., Fajardo A.R., Rubira A.F., and Muniz E.C. 2012. Superabsorbent hydrogel nanocomposites based on starch-g-poly (sodium acrylate) matrix filled with cellulose nanowhiskers. *Cellulose,* 19(4):1225–1237.

[36] Cavaille J, Chanzy H, Fleury E, and Sassi J. 2000. Surface-Modified Cellulose Microfibrils, Method for Making the Same, and Use Thereof as a Filler in Composite Materials. *U.S. Patent No. 6*, 117, 545.

[37] Heux L., Dinand E., and Vignon M.R. 1999. Structural Aspects in Ultrathin Cellulose Microfibrils Followed by 13C CP-MAS NMR. *Carbohydrate Polymers*, 40:115–124.

[38] Schroers M., Kokil A., and Weder C. 2004. Solid Polymer Electrolytes Based on Nanocomposites of Ethylene Oxide-Epichlorohydrin Copolymers and Cellulose Whiskers. *Journal of Applied Polymer Science*, 93:2883–2888.

[39] Orts W.J., Shey J., Imam H.M., Glenn M.G., Gutman E.M., and Revol F. J. Application of Cellulose Microfibrils in Polymer Nanocomposites. *Journal of Polymers and the Environment*, 13(4), 301–306 (2005).

[40] Helbert W., Cavaille J.Y., and Dufresne A. 1996. Thermoplastic Nanocomposites Filled With Wheat Straw Cellulose Whiskers. Part I:

Processing and Mechanical Behavior. *Polymer Composites*, 17(4):604–611.

[41] Kallel F., Bettaieb F., Khiari R., García A., Bras J., and Chaabouni S.E. 2016. Isolation and structural characterization of cellulose nanocrystals extracted from garlic straw residues. *Industrial Crops and Products*, 87:287–296.

[42] Fleming K., Gray G.D., Prasannan S, and Matthews S. 2000. Cellulose Crystallites: A New and Robust Liquid Crystalline Medium for the Measurement of Residual Dipolar Couplings. *Journal of the American Chemical Society*, 122(21):5224–5225.

[43] Dong X.M., Kimura T., Revol J. F., and Gray D.G. 1996. Effects of Ionic Strength on the Isotropic-Chiral Nematic Phase Transition of Suspensions of Cellulose Crystallites. *Langmuir*, 12(8):2076–2082.

[44] Favier V., Chanzy H., and Cavaille Y.J. 1995. Polymer Nanocomposites Reinforced by Cellulose Whiskers. *Macromolecules*, 28:6365–6367.

[45] Akhlaghi P.S., Berry C.R., Tam C.K. 2013. Surface modification of cellulose nanocrystal with chitosan oligosaccharide for drug delivery applications. *Cellulose*, 20:1747–1764.

[46] Yan H., Chen X., Feng M., Shi Z., Zhang W., Wang Y et al. 2019. Entrapment of bacterial cellulose nanocrystals stabilized Pickering emulsions droplets in alginate beads for hydrophobic drug delivery. *Colloids and Surfaces B: Biointerfaces*, 177:112–120.

[47] Lim S.L., Rosli A.N., Ahmad I., Lazim M.A., and Amin M.I.C.M. 2017. Synthesis and Swelling Behavior of pH-Sensitive Semi-IPN Superabsorbent Hydrogels Based on Poly (acrylic acid) Reinforced with Cellulose Nanocrystals. *Nanomaterials*, *7*:399.

[48] Okamoto T., Patil J.A., Nissinen T., and Mann S. 2017. Self-Assembly of Colloidal Nanocomposite Hydrogels Using 1D Cellulose Nanocrystals and 2D Exfoliated Organoclay Layers. *Gels*, *3*:11.

[49] Supramaniam A., Adnan R., Kaus M.H.N., Bushra R. 2018. Magnetic nanocellulose alginate hydrogel beads as potential drug delivery system. *International Journal of Biological Macromolecules*, 118:640-648.

[50] Villanova J.C.O., Ayres E., Carvalho S.M., Patrício P.S., Pereira F.V., Oréfice R.L. 2011. Pharmaceutical acrylic beads obtained by suspension polymerization containing cellulose nanowhiskers as excipient for drug delivery. *European Journal of Pharmaceutical Sciences,* 42:406–415.

[51] Guo R., Lan Y., Xue W., Cheng B., Zhang Y., Wang C., and Ramakrishna S. 2017. Collagen-cellulose nanocrystal scaffolds containing curcumin-loaded microspheres on infected full-thickness burns repair. *Journal of Tissue Engineering and Regenerative Medicine*, 11:3544–3555.

[52] Lin N., and Dufresne A. 2013. Supramolecular Hydrogels from In Situ Host−Guest Inclusion between Chemically Modified Cellulose Nanocrystals and Cyclodextrin. *Biomacromolecules*, 14, 871−880.

[53] Hasan A., Waibhaw G., Tiwari S., Dharmalingam K., Shukla I., Pandey M.L. 2017. Fabrication and characterization of chitosan, polyvinylpyrrolidone, and cellulose nanowhiskers nanocomposite films for wound healing drug delivery application. *Journal of Biomedical Materials Research Part A*, 105A:2391–2404.

[54] Taheri A., Mohammadi M. 2015. The Use of Cellulose Nanocrystals for Potential Application in Topical Delivery of Hydroquinone. *Chemical Biology & Drug Design*, 86:102–106.

[55] Dua H., Liu W., Zhang M., Sia C., Zhang X., Li B. 2019. Cellulose nanocrystals and cellulose nanofibrils based hydrogels for biomedical applications. *Carbohydrate Polymers* 209:130–144.

[56] Karimian A., Parsian H., Majidinia M., Rahimi M., Mir M., Smadi-Kafil H., ShafieiIrannejad V., Kheyrollah M., Ostadi H., Yousefi B. 2019. Nanocrystalline cellulose: Preparation, physicochemical properties, and applications in drug delivery systems. *International Journal of Biological Macromolecules,* 133:850-859.

[57] Klemm D., Cranston D.E., Fischer D., Gama M., Kedzior A.S., Kralisch D., Kramer F., Kondo T., Lindström T., Nietzsche S., Petzold-Welcke K., Rauchfuß F. 2018. Nanocellulose as a natural source for groundbreaking applications in materials science. *Today's state. Materials Today* 21:720-748.

[58] George J., Sabapathi N.S. 2015. Cellulose nanocrystals: synthesis, functional properties, and applications. *Nanotechnology, Science and Applications.* 8:45–54.

[59] Hamad Y.W., Atifi S., Berry M.R. 2014. *Flexible nanocrystalline cellulose (ncc) films with tunable optical and mechanical properties* WO 2014/138976 Al.

[60] Jahan Z, Niazi K.B.M., Gregersen W.Ø. 2018. Mechanical, thermal and swelling properties of Cellulose Nanocrystals/PVA nanocomposites membranes. *Journal of Industrial and Engineering Chemistry,* 57:113-124.

[61] Atifi S., Su S., and Hamad Y.W. 2018. Mechanically tunable nanocomposite hydrogels based on functionalized cellulose nanocrystals. Nordic *Pulp & Paper Research Journal*, l:29 (1).

[62] Thomas D., Latha M.S., and Thomas K.K. 2018. Synthesis and in vitro evaluation of alginate-cellulose nanocrystal hybrid nanoparticles for the controlled oral delivery of rifampicin. *Journal of Drug Delivery Science and Technology*, 46:392–399.

[63] Sarkar A., Zhang S., Murray B., Russell A.J., Boxal S. 2017. Modulating in vitro gastric digestion of emulsions using composite whey protein-cellulose nanocrystal interfaces. *Colloids and Surfaces B: Biointerfaces*, 158:137-146.

[64] Low E.L., Tan H. T.L., Goh B. H., Tey T.B., Ong H.B., Tang Y.S. 2019. Magnetic cellulose nanocrystal stabilized Pickering emulsions for enhanced bioactive release and human colon cancer. *International Journal of Biological Macromolecules*, 127:76-84.

[65] Mauricio R.M., da Costa G.P., Haraguchi K.S., Guilherme R.M., Muniz C.E., Rubira F.A. 2015. Synthesis of a microhydrogel composite from cellulose nanowhiskers and starch for drug delivery. *Carbohydrate Polymers*, 115:715–722.

[66] Bajpai S.K., Chand N, and Ahuja S. 2015. Investigation of curcumin release from chitosan/cellulose microcrystals (CMC) antimicrobial films. *International Journal of Biological Macromolecules* 79:440–448.

[67] Bourne D.W.: Pharmacokinetics. In: *Modern Pharmaceutics*. 4th ed. Banker GS, Rhodes CT, Eds., Marcel Dekker Inc, New York, 2002.

[68] Grassi M., Grassi G. 2005. Mathematical modeling and controlled drug delivery: matrix systems. *Current Drug Delivery*, 2(1):97-116.

[69] Higuchi T. 1963. Mechanism of sustained action medication. Theoretical analysis of rate of release of solid drugs dispersed in solid matrices. *Journal of Pharmaceutical Science*, 84:1464.

[70] Shoaib H.M., Tazeen J., Merchant A.H., Yousuf I.R. 2006. Evaluation of drug release kinetics from ibuprofen matrix tablets using HPMC. *Pakistan Journal of Pharmaceutical Sciences,* 19:119.

[71] Rees A, Powell C.L., Chinga-Carrasco G., Gethin T.D., Syverud K, Hill E.K., and Thomas W.D. 2015. 3D Bioprinting of Carboxymethylated-Periodate Oxidized Nanocellulose Constructs for Wound Dressing Applications. *BioMed Research International*, 925757, 7.

[72] Panchagunla R. 1997. Transdermal delivery of drugs. *Indian Journal of Pharmacology,* 29:140-157.

[73] Neubert R.H.H. 2011. Potentials of new nanocarrier for dermal and transdermal drug delivery. *European Journal of Pharmaceutics and Biopharmaceutics*, 77:1-2.

[74] Patel D., Chaudhary A. S., Parmar B., Bhura N. 2012. Transdermal drug delivery: A review. *The pharma innovation,* 1:4.

[75] Xu Q., Ji Y., Sun Q., Fu Y., Xu Y., Jin L. 2019. Fabrication of cellulose nanocrystal/chitosan Hydrogel for controlled drug release. *Nanomaterials* 9:253.

[76] *Emerging Trends in Science and Technology; Acid Recovery Based Novel Route for the Synthesis of Nanocellulose from Lignocellulosic and Cellulosic Fibres*. Authors: ISBN: 978-81-8487-642-0, E-ISBN: Publication Year: 2019, 248, Binding: Hard Back, Dimension: 160mm x 240mm. Authors: Orasugh T. J., Basu A., Ghosh K. S., Chattopadhyay D. Editor(s): Ganguly K., Kumar A., Chakrabarti S. Narosa Publishing House.

[77] Orasugh T. J., Dutta S., Das D., Pal C., Zaman A., Das S., Dutta K., Banerjee R., Ghosh K. S., Chattopadhyay D. 2019. Sustained release

of ketorolac tromethamine from poloxamer 407/cellulose nanofibrils graft nanocollagen based ophthalmic formulations. *International Journal of Biological Macromolecules*, 140, 1, 441-453.

[78] Orasugh T. J., Dutta S., Das D., Nath N., Pal C., Chattopadhyay D. 2019. Utilization of Cellulose Nanocrystals (CNC) Biopolymer Nanocomposites in Ophthalmic Drug Delivery System (ODDS). *Journal of Nanotechnology Research*, 1 (2): 075-087.

INDEX

#

A

B

D

E

F

G

H

I

K

L

M

S

T

Related Nova Publications

First-Principle vs. Experimental Design of Diluted Magnetic Semiconductors

Authors: Omar Mounkachi, Abdelilah Benyoussef, and Mohamed Hamedoun

Series: Nanotechnology Science and Technology

Book Description: The purpose of this book is to propose some ideas to answer the most important question in material science for semiconductor spintronics, primarily considering how room-temperature ferromagnetism in DMS can be realized. Additionally, the correlation between first principle and experimental design to see how properties of yet-to-be-synthesized materials can be predicted is discussed.

Hardcover ISBN: 978-1-53614-077-4
Retail Price: $195